I0461905

The circle of underachievement

By

Ian Jordan

Acknowledgements

There have been a great number of people who have helped me in many ways and I have learnt so much from them that I can only say this book could not have been written without them. I must thank those who have had such faith in me, even when I didn't have as much faith in myself, in particular, Dr Graham Street for his advice and help, Ian and Carol Hailes, Clive Murray, Graham Rennie and Vince, Fred and Lindsay Webster. I would also like to thank "First impressions" of Portsoy, Frank Norville, Paul Gibbons and Tom Jones of the Norville Group, Steve Ellis from Hoya, Peter Reeves, John Newth from CTP COIL, Elizabeth Fraser, Dr Beverley Steffert, John Anderson, Paul Wightson, Prof Yuri Kropotov, Prof R Craz, Dawn Deacon and her "bunnies" from HMP Gartree, Carol Parkes from Laxton Prep School, Mary Evans from Gordonstoun School, the staff and patients of I.C. Andrew Opticians, Steve Laws of Desktop Publications, my sons, David and Paul for the enormous help they have given in experiments, and my long suffering and patient wife, Beatrice, who has contributed more to this book than she will ever admit.
Thank you.

Contents

Appendices

Preface

There is an enormous amount of research being undertaken into the visual aspects of reading difficulties and dyspraxia, but there is little information available to the non-professional. The technical books that have been published are generally found unreadable by the layman. They are often contradictory and, consequently, it is extremely difficult for parents to be able to recognize whether visual dyslexia is present in their child. The technical books do not give a parent the information they require on assessment and treatment on how to achieve the best results. This book is not designed to be a highly technical treatise and some ideas are unproven but, in my opinion, they are the best explanations, at present.

This book gives a balanced view of current assessment methods and the treatment available and will help many to become aware of the importance of addressing the physical disabilities and perceptual problems associated with dyslexia. It introduces a number of new tests and a significant amount of background information. Mathematical principles of binocular vision are explained simply and the connection, with a number of physical conditions, will be explained.

The relationship between visual perception and hearing and balance is explored and how this affects dyspraxia and dyslexia. Some of the visual phenomena that are described and can be demonstrated, inevitably, call into question some of the accepted "science" of dyslexia. That is a good thing, as we only move forward by extending and questioning the status quo. After all, it is only recently that, due to pioneers fighting to change attitudes, dyslexia has become a recognised condition.

My web site is www.visualdyslexia.com and I may be contacted by a link from that site. I am sorry that I will not be able to comment on individual cases, unless I have assessed them. Tests described in the book are available via links from the website.

A second website www.orthoscopics.com will be of help to those that require information on assessment using techniques described in the book.

Any mistakes omissions or inaccuracies are my responsibility, as are any conclusions drawn. Please make this book obsolete by improving on the work detailed inside.

Chapter 1

Introduction

"The circle of under-achievement " was conceived as a result of the amount of interest shown by parents, teachers and other professionals in the reading problems, exhibited by so many, that cannot be explained by language difficulties alone. On my lecture tours, there is a recurring demand for an accessible book that would provide a rounded view of how visual perception can influence virtually every activity that is undertaken. There are a number of extremely complex scientific papers and books on aspects of visual dyslexia and other perceptual problems, but there is relatively little for the enquiring mind of the lay person.

This book will, I hope, paint a broad picture of this complex area, explain the major theories, in understandable terms, and allow a balanced view to be formed by a parent or sufferer.

We use the term visual dyslexia to describe the group of symptoms that are common to a number of perceptual problems but always cause difficulties with reading text. These will be discussed in detail later but can be described simply as unstable vision. This book will use the term visual dyslexia to differentiate visually evoked symptoms rather than cognitive responses.

There are many new ideas and tests described in this book (some for the first time) all of which have scientific backing to a greater or lesser extent although some may not have been through clinical trials in the form in which they are presented. This will be made clear in the book.

Although many of the techniques, which I describe, are innovative, they have all been demonstrated publicly to peer-group audiences and any resulting criticisms are noted in the text. Many of the methods are very exciting and the results can be dramatic.

This book is written primarily for parents who suspect that their children may have problems, but it applies equally to adults who experience similar difficulties.

A parent's role

What do parents do when their child is not achieving his or her educational / reading potential?

Do they accept the situation or do they investigate the cause (or causes) and then take action where possible? Most parents will react by trying to do their best with the resources available, but, where educational difficulties are concerned, they will often have great difficulty in accessing the best possible advice and support. So, what is the best course of action?

Sadly there is NOT a one-stop-shop for reading problems and therefore some knowledge is required to access appropriate help. Dyslexia is only one of a number of possible causes of difficulty and a high level of skill is required to differentiate between them

Is a visual perceptual problem present?

Do NOT rely on your child's teacher to give the best advice. Even though you would assume the class teacher would have significant levels of expertise with visual dyslexia, it is a fact that many have little or no training in recognition techniques and their advice is idiosyncratic as a result. It is unreasonable to expect them to be expert without training. You may have to be your child's champion.

Visual dyslexia?

In order to treat visual dyslexia successfully, it is necessary to comprehend how the physical aspects of perceptual difficulties interact. It is useful to be able to understand the reactions of the professionals and be aware of how they reach a diagnosis and prognosis. It is imperative that the disparate nature of visual perceptual difficulties be understood and that it become apparent that there is not one condition but a number of problems that are collectively named. It follows that the way forward must reflect and treat the different symptoms encountered, although there is often an overlap of techniques.

There are a number of assessment techniques available; this book should give a basic understanding of the these. Treatments will be explored in some detail. Possible treatments vary and more than one treatment regime can be helpful. This book will give enough information for an informed consent and will explain why different approaches have validity.

The parent will often go to one or a number of other professionals to seek advice. This advice may be contradictory, expensive and may promise miracles. Suggestions of other treatments will possibly include visual, auditory, dietary, educational, psychological, kinesthetic, manipulation, biofeedback, drugs etc.

All treatments (even placebo) may be useful for some individuals. It may be that more than one method of treatment will achieve satisfactory results and, then, it may be a matter of informed choice as to which method is chosen as the most applicable.

Conflicting advice

As soon as a child is suspected to have a difficulty, many people will generously advise the parent on how to proceed. This advice is often contradictory and sometimes damaging to the child. It is often said "they will grow out of it" and inaction is the result. This may be very bad advice and may be influenced by financial considerations, particularly if a school has to find money from an already stretched budget to pay for assessment and treatment.

So, how does a parent respond when they are given so much conflicting advice?

The parent is the ultimate professional - use your knowledge and common sense.

Choosing a professional

It is said that there is no method for the parent to evaluate the standard of care given. Professional qualifications may be misleading, as many professionals in related fields have limited knowledge or training in areas associated with reading difficulties. Common sense should be used by the parent in deciding on the veracity and ability of the practitioner.

The assessment for visual dyslexia may be long, involved and, sometimes, costly. Some practitioners have limited knowledge and may do significant harm giving incorrect advice. It is possible to "treat" dyslexia of any type without any training or study. There are however, a significant number of practitioners who are trying to assimilate the various types of knowledge required enabling them to give a good level of care.

Ask the right questions.

This book should help the parent or teacher to pose the right questions about the visual aspects of reading difficulties and to approach the subject scientifically, rather than intuitively. It is only recently that the visual aspects of dyslexia have become measurable and, now, it is no longer satisfactory for the professional to ignore this advance. After all, at least 10% of people have the visual problems associated with dyslexia!

Is visual dyslexia a psychological problem?

At present, dyslexia and other reading difficulties are assumed to be a cognitive problem that is best dealt with in a psychological way. Whilst accepting that processing problems may result in some of the symptoms of dyslexia it is now apparent that physical processes are of vital importance in the majority of children with reading problems and these MUST be addressed, prior to educational or psychological intervention.

Not to address these physical difficulties is unacceptable, as all further interventions may be flawed. The question must be asked "if a child is not achieving his potential, who should I see first?"

The answer must be - the person who will alleviate the physical problems, whether they are visual, auditory or proprioceptive.

Professionals and dyslexia

An ideal assessment for a child, that is underachieving, would involve consultation with a number of professionals. These may include:-

      Optical professionals

      Auditory professionals

      Occupational therapists

      Dietary professionals

      Educational psychologists

      Teachers

      Chiropractors

What can I expect from my school?

Often, the parent is confused as to why their child is not reading as well as they would anticipate and, often, the class teacher is concerned at the lack of progress of one of their children. It is the responsibility of the teacher to spot the difficulties that the child is experiencing and to take appropriate action. Quite a feat, with a class of 30! Large class sizes will often result in the teacher being unable to spot a child that is having problems.

Often, the training of teachers for children with perceptual difficulties is limited (less than a day of the teacher training course may be devoted to special needs) and, frequently, financial and time pressures are brought to bear, thereby ensuring an inadequate response. It often appears to me, that, at this point, societies (or our political masters) demonstrate clearly how little they value the CHILD with problems. Contrast this with the large sums given to universities and further education colleges for their students with dyslexia. Is this because the adult student, particularly one with ability to access higher education, will not accept inadequate provision?

The most important intervention is early intervention - do not accept shoddy treatment.

Diagnosis

It is unreasonable to expect the class teacher to diagnose and be responsible for physical intervention; they have to recognize that the problem exists and refer the child on to the correct person, who can deal with the problems that it is exhibiting. At present, this referral is, usually, to an educational psychologist.

There are inherent problems with this approach, as physical treatments are often neglected at the expense of educational treatments, and consequently optimum results are not achieved in many cases. I would suggest that it is better that teachers have a duty to inform the parent of the child if they suspect that the child has a problem. It would then be the responsibility of the parent or person with parental responsibilities to ensure that their child was assessed and treated appropriately.

The idea that teachers and psychologists alone should be responsible for intervention is worrying, and has proved

17

unsatisfactory for many. There is an obvious role for the teacher and psychologist, subsequent to physical assessment and intervention. It is more efficient for them to be dealing with genuine educational and psychological problems than physical difficulties.

Optical assessment

Optometrists, ophthalmologists, opticians and orthoptists may have some knowledge of visual dyslexia. It is not widely known that their expertise can vary wildly and there is a wide spectrum of ability, ranging from having virtually no knowledge to being expert. It is very difficult for a member of the public to determine the levels of knowledge and experience, but the local optical "grapevine" can be accessed from your local optician. Hospitals are rarely the place where visual perceptual assessments are undertaken and the health service, at present, will generally not pay for them.

An eye test is an inadequate test for visual dyslexia and will often result in the statement from the optometrist "there is nothing wrong with the child" even when there are significant perceptual difficulties. This should be re-phrased to "there is an insignificant refractive error, but you may have a visual perceptual problem" and "either I can perform the assessment necessary or I can refer you to another practitioner, who is able to perform all the necessary tests". You will be pleased to know that the optical professions are moving towards higher standards and this can only be encouraged.

The dyslexia industry

There is an enormous dyslexia industry that may be unhappy with the idea that a significant proportion of sufferers can be treated extremely successfully and predictably. Sadly, in the past, many commercial interests, which purported to help dyslexics, helped themselves first and may have at times, hindered progress. They may have charged large sums for minimal work and inadequate treatments. These costs, for some children from poorer families, made it impossible to have visual dyslexia diagnosed!

However, a well-qualified person with good equipment and sufficient time to provide an adequate assessment will cost money. Eventually, someone has to pay for assessment; underfunding will mean that corners are cut. Serious concerns as to the veracity and accuracy of educational tests may be alleviated by ensuring that the assessor is independent, as pressure can and is brought to bear on both teachers and educational psychologists by their employers.

Educational policies

Visual dyslexia is not directly related to intelligence, although a child with a high IQ and low ability will be more obvious and therefore often diagnosed earlier. Sadly, many of lower ability are left to vegetate, as they are not expected to achieve. This lack of expectation becomes a self-fulfilling prophecy. They do not improve! Most can and it is often more critical that those of limited ability get the maximum help. Some local authorities set a percentage of children that they consider to have special needs; they may say the lowest 1% of achievers will receive extra help. This position is disgraceful and unjustified – the special needs of an area can only be judged on the amount of children that require extra help, not set to a figure and manipulated so that the children fit it, arbitrarily. Teachers or educational psychologists should be asked to relay the policy of the educational authority, in their area. Terms of employment could prevent the educational psychologist or teacher from talking or helping the parent or child by gagging them, even when it is in the best interests of the child to discuss matters. Whilst there are good reasons why any employee should sometimes be prevented from discussing work issues, it is imperative that the professional educationalists should recognise, that the interest of the child is paramount and that they must be free to use their judgement.

There are a number of methods which a parent can use to access professional help. These include access through the school, writing to the director of education or paying privately. Parent partnership may also be able to help – your LEA will tell you how. Do not forget that legal help is available and that, usually, the child will benefit from Legal Aid. Damages can be sought and can be considerable, in extreme cases. It is not unusual for significant obstacles to be placed in front of the parent (it varies widely, depending on postcode). Be persistent. The vast majority of those that undergo assessment and treatment are from families that know how to use the system; this usually means children from a well educated background, with eloquent and assertive parents. They can afford the assessments and have a higher expectation.

It is a matter for society to protect those children without parents who can access the best treatment; we all have a responsibility. Society cannot afford the segregation of those who cannot afford treatment; could it be the reason why such a high proportion of juveniles in prison are dyslexic? If many children are not perceived as being worth helping, it is inevitable that antisocial behaviour will be the consequence. Why is a truant so often a dyslexic? Self worth is often reduced for the dyslexic child in this literate society and we must insist that all children get appropriate help.

It must be stated that only a small percentage of those with dyslexia are antisocial. The rest get on with life, without causing problems, by developing coping strategies, although they cannot achieve their potential as easily as those without problems. It is much more difficult for a dyslexic child to achieve academic success. Those that do, work harder!

Costs

The cost of not helping a dyslexic child may be far higher than the cost of helping. I suspect that it would cost at least ten times as much to society by not taking appropriate action by assessing and treating every dyslexic child, at the earliest time. Society has to live with the inevitable consequences.

This is a long-term ideal; perhaps a far-seeing politician may take a longer-term view than the next election!

The science of visual dyslexia and perceptual treatment

The science of visual dyslexia is new, fascinating, and changing rapidly. It is tremendously exciting for me to see the way, in which the scientific inter-relationships are developing. In this context, I will describe the Cambridge trials that we have just completed and how they should influence future assessments and treatments for all those with reading difficulties.

Chapter 2

Anatomy

This chapter deals with the anatomical and physical structures of the visual system in a simple way, to enable non-clinicians to understand basic anatomical and neurological concepts. We will explore physiological problems in greater detail, later. This book is not designed as a clinical textbook and therefore we will only describe the structures that are necessary for an understanding of the visual aspects of dyslexia. We will deal with the eye, the brain and the ear, concentrating principally on the visual pathways, the anatomical and physiological aspects of the eye and the way that these integrate with the ear and the balance systems. Knowledge of these systems is essential to understand the processes present in the visual aspects of dyslexia.

The anatomy of the visual system is complex and I make no apology for simplifying it. I appreciate that many of the words are effectively a foreign language. Nerve pathways in particular have been reduced to a minimalist level, as their anatomy is daunting to all but the most dedicated. It is however crucial to have a basic understanding of these pathways for comprehension of some of the principles in visual dyslexia.

The eye and visual system

The visual system is extraordinarily complex and is not yet fully understood.
It consists of the eye and connections to and from the brain. It interacts with the auditory (hearing) and vestibular (spatial awareness/balance) systems. Developmental processes affect it, as well as disease and changes in body chemistry. Environmental factors modify its use. Vision is a superb and, to most of us, the most wonderful of the senses.

The eye develops in the womb from modified brain tissue. It is an opaque orb with a variable aperture (pupil) that allows light to enter. The light then passes through a variable focus lens that inverts the image on the retina. The light energy is converted to a signal by chemical processes in the retinal receptor cells. This signal travels along nerves to the brain, which then analyses the information and either responds automatically, discards the information or allows a decision, to respond or not, to be taken consciously.

Visual input

When a person looks into the distance, both eyes see a very similar but slightly different picture. The nearer an object, the greater this difference becomes. The light from the viewed object enters the eye through a clear structure (cornea) that is steeply curved and bends the light to a focus inside the eye. This light passes through the pupil that becomes larger (dilates) and smaller (constricts) as a response to light intensity and as the eye focuses (accommodates) on a nearer object. This change in pupil size regulates the amount of light entering the eye and changes the depth of focus. The light is focused in the clear jelly-like area (vitreous), and inverts the image received on the light sensitive back of the eye (retina).

Rods and cones

The cells in the retina are divided into two types, rods and cones. The rods are sensitive in low light conditions and do not differentiate colour, the cones require higher energy i.e. more light to cause firing and are colour sensitive. They form a jigsaw or mosaic pattern on the retina, with the cones at their peak density in the central vision area (fovea) reducing in the peripheral retina. The rods are predominantly found in the peripheral areas of the retina.

Colour sensitivity is trichromatic i.e. all perceived colours can be built up using the three primary colours. There are three forms of cones that have sensitivities that peak in the red, green and blue wavelengths of light. The cells within the retina undergo a photochemical change stimulated by the light energy and "fire". This produces a nerve impulse that travels along a nerve pathway to the brain.

Magnocellular and Parvocellular pathways

Cone cells appear to have at least two separate major pathways, the Magnocellular ("M") and Parvocellular ("P"). These appear to have different tasks. The M pathway responds to edges and flicker, the P to form. It is now believed that, as well as having excitation effects, these may also inhibit the inputs of each other.

M cells respond to edges e.g. black and white borders, by firing, when the stimulus changes. Constant monochromatic stimulus results in no firing, after the first initial discharge. P cells respond to form and texture.

These systems have different characteristics that respond, on the one hand, to the transient and, on the other, to the sustained visual stimuli and these are combined within the brain, subconsciously. The easiest way, to think of these two systems combining, is like a cinema film, that produces a perceived constant moving image from twenty-five still images per second. It is likely that the magnocellular pathway has a significant effect on eye movement control and that this has led to speculation that it is dominant in visual dyslexia.

Functional development of vision

The firing of retinal cells develops in a baby, as a response to changing visual stimulation. This involves both light and form. If this stimulation is not adequate e.g. in congenital cataract, the results may significantly influence visual development. In extreme cases, visual function may not be present and blindness may be the consequence. Visual function is a learned response and inadequate stimulation may result in poor vision, due to poor visual processing. Suppression of the visual stimulus is also learnt, as unwanted images will confuse the visual system. This reaction to stimuli is of major significance in visual dyslexia, as well as in amblyopia (lazy eye) and strabismus (squint).

The Optic Nerves

I will summarize the principal nerve pathways in such a way, that they can be understood easily. In the retina, the rods and cones activate bipolar cells, also in the retina. At either end of the bipolar cells there are cells (horizontal and amycrine) that are connected in a spiders-web like pattern. This allows one cell to be de-sequenced or laterally inhibit (or excite) adjacent cells and areas. I believe this separation to be of great importance in visual dyslexia, as it will allow an explanation for many of the symptoms experienced.

The horizontal and amycrine cells connect to large ganglion cells which combine and emerge from the back of the eyeball as the optic nerve. From the eyeball this nerve enters the cranial fossa and passes through the chiasma where the nasal fibres from each eye cross over to fuse with the temporal fibres of the other eye. The optic tract then passes to one of three destinations; the lateral geniculate body for relay to the visual cortex, the pretectile nuclei which governs pupil reactions to light, and the superior caliculus for body reflexes to light. There are a number of nerves that appear to be effected by visual input; these include the third nerve, the fourth nerve, the seventh nerve and perhaps the most important for symptoms, the fifth nerve.

The trigeminal nerve (the fifth cranial nerve) has a number of functions. It governs tear production, it operates the upper eyelid and it carries hitchhiking sympathetic fibres to the dilator muscle of the iris. It also innervates the tensor tympani muscle which controls the tonus of the ear drum. It is implicated in headaches and migraine.

We will return to the functional differences later.

Retinal correspondence

The signals from the nerves from the nasal side of each retina cross (at the chiasma) to the opposite side of the brain. There the transmitted information has to be integrated or fused with the information sent from the temporal retina of the opposite eye. For a distant object both eyes receive almost identical images and the correspondence of the images on the retina is close. The two images are relatively easy to fuse because a certain tolerance of corresponding fusional areas must be available. This fusion gives the effect of three-dimensional vision. Retinal fusional areas may play a big part in dyslexia and are the basis of the new model that I will suggest, later in the book.

 However during convergence there is often a marked difference in the images in the corresponding areas of the retinas. Inevitable double vision occurs (diplopia) that has to be dealt with by the visual system. The information may be suppressed either partly or fully from either or both eyes. This suppression may be inconsistent or alternating and may be the cause of many problems, such as some forms of squint. It may also be one of the principal causes of visual dyslexia.

In reading to the right of the mid line there are positions where a crossover of the temporal image of the left eye moves from the right side of the mid line to the left of the mid line. It is impossible for the brain to merge or fuse these significantly different images. Depending on position of gaze, the distance from the eyes, the size of the words in the text and the width of the interpupillary (distance between the eyes) distance there can be a number of effects:

- Word displacement
- Reversal
- Doubling or shadowing

There is also a major change in relative image size due to traversal of text that may have profound effects on fusion and fixation. Accommodative effort varies between the eyes and aberrations switch from one eye to the other. Sloping of images is inevitable with the direction of slope mirrored between the image perceived in each eye.

What options does the brain have?

- Double vision on convergence

- Suppression of the central vision in either the left or right eye and rotation of an eye to enable fusion and stabilization of the peripheral fields.
- Suppression of one of the peripheral fields
- Incomplete or variable suppression
- Complete suppression of one eye

It is apparent that the brain can and does take all these options, at some time, in some individuals.

Anatomical structures play a large part in visual dyslexia.

Fixation Development

At the age of five it is estimated that about 50% of all children do not have the ability to maintain fixation on a word in the near field. At ten years of age this reduces to about 5%. This causes significant difficulties in reading development. Fixation must be learnt. Failure to be able to fix on a word is certainly one of the other causes of visual dyslexia. It is normal for a young child to have some fixation problems, when starting to read, but abnormal if fixation does not develop quickly.

Mechanics of the eye

When looking at a close object, it is necessary for the eyes to adjust the focusing power of the lens within the eye, and also to rotate the eyes towards the nose, to allow alignment of vision and to suppress the inevitable double vision that is produced on rotation. (This suppression may be demonstrated by viewing a finger at 25cm and closing each eye in turn. The suppressed image will appear to jump into place.)

Around the eyes are six muscles which rotate the eyes to a fixation point, during reading and move the eyes along the line in short jerky movements called visual saccades. These saccades consist of fixation of the eyes on a word, followed by a jump to a word further along the line, with small corrective and reverse movements to achieve the best possible fixation position and focus. Smooth eye movements also occur e.g. tracking a moving ball such as a tennis ball (pursuit movements) . This system develops in childhood and is almost invariably affected in a child who shows magnocellular symptoms. Slight rotary movements are also possible e.g. as a response to head rotation (the vestibular reflex). Movement control of the eyes may also be implicated in some with

35

dyslexic symptoms and may be treated by exercises or by modification of the magnocellular input. The parietal lobe and the cerebellum are crucial in saccadal movement control in reading.

The eyes may be emmetropic (focus perfectly), be long sighted (hyperopic), short sighted (myopic) or astigmatic (having a different regular shape like a rugby ball). These conditions are due to the length and shape of the eyeball. Although they do not directly affect visual dyslexia they do have a number of indirect effects.

The Auditory System

The auditory system consists of three distinct areas, the outer ear, the inner ear and the middle ear. The outer ear works as a receptor for the sound waves to be deflected into the middle ear. Sound waves cause the eardrum to vibrate this vibration is passed into the inner ear via three small bones, the malleus, the incus and the stapes. These cause the fluid within the cochlea to vibrate. This vibration moves very small hairs called cilia and these respond by sending a signal to the brain. The cochlea is continuous, with three fluid filled cavities called the semicircular canals. Movement causes the fluid to move against cilia causing innervation that allows a rotational positional sense to be established. Balance is thereby partly determined. The balance system within the semicircular canals appears to be a modified auditory system. The balance system is also dependant on the position and rotation of the eyes, as they show compensatory actions during head rotation. Proprioceptive cells, which are found all over the body, send signals to the brain telling the brain their orientation and position and this allows the position of the body to be assessed. This combined system is called the vestibular system. It is interesting to note that in some people signals from the visual system seem to over-ride the signals from the ears, resulting in a tremendous lack of balance when the eyes are closed. In some people with both eyes and ears working it is possible to improve their balance and hearing with visual treatment and it is likely that auditory treatment may be of assistance in visual dyslexia. Extremely strong auditory or vestibular stimulation may result in nystagmoid (a jerky movement similar to saccadal) eye movement or changes in pupil size. Other disturbances of the vestibular system e.g. weightlessness in space may also produce these unpleasant effects

There are two muscles within the ear, which are of particular interest in dyslexia, the tensor tympani and the stapedius. The tensor timpani occupies the canal above the bony part of the auditory tube. Contraction of this muscle draws the handle of the malleus bone inwards, making the eardrum more highly concave and therefore tenser. Its nerve supply is from the mandibular nerve (a branch of the trigeminal); the stapedius arises from the interior of the hollow pyramid, it is inserted into the neck of the stapes and retracts the neck of the stapes thus tilting the foot piece in the oval window. Paralysis of this muscle causes an abnormal increase in the power of hearing (hyperacuisis). It is believed that this muscle may play a part in filtering the background sounds. It is supplied by the facial nerve.

Auditory and vestibular integration.

As the visual nerve pathways proceed towards the brain there are three positions in which they cross with the auditory /

vestibular nerve pathways. There are a number of parallels of innervation between the two systems and some cells in the Magnocellular system are innervated by both auditory and visual input.

Genetics

It has become apparent that the visual form of dyslexia runs in families. It is very likely that, if one child has problem, other children in the family will also suffer from the same or associated disability. They often produce very similar responses to testing techniques, although it is likely that one will have greater magnitude of response. Indeed it is almost obligatory to test other members of the family if one member shows symptoms. Research has shown that there is a genetic influence on many of the symptoms and a number of chromosomes have been implicated, but the evidence is not conclusive. These chromosomes are close to chromosomes that are believed responsible for atopic (allergic) conditions such as asthma.

The brain

The anatomical structure of the brain is complex and not within the scope of this book. There are some functional aspects of the brain that have an influence on visual dyslexia and therefore they must be addressed. The brain is generally considered as a neural network in which nerve cells communicate. Although we normally talk of the right and left hemispheres being sent the messages uniquely from particular cells it appears likely that this is too simplistic. The right and left hemispheres "talk" to each other via the corpus collosum (which is much larger in females). The receptive fields typified in the retina continue throughout the nerve pathways to the visual cortex via the lateral geniculate nucleus. Speech is converted from visual input at the angular gyrus.

Where is visual dyslexia produced?

There are three options:

Input dysfunction i.e. the signals sent from receptors is aberrant

Nerve pathway dysfunction i.e. the signals are transferred incorrectly to the brain or the signal is distorted

Response dysfunction i.e. the brain responds incorrectly

Each of these appears to be a possibility in some subjects with visual dyslexia.

The question must be posed "is dyslexia a physical problem?"

The answer must be "generally yes"

Chapter 3

Visual organisation

Introduction

In this chapter we will deal with the mechanics of reading a block of text.

It is an extremely complex process and is still not understood fully. Frankly, I doubt that it ever will be!

There are many opinions in respect of the mechanisms involved in reading. Current research is moving frontiers, particularly in the understanding of information processing.

There are a number of steps necessary to process visual information

- The input signal to the brain must be adequate
- The connections within the brain must allow for correct interpretation of the signal from the eye
- The brain should filter the information to allow important information to be analysed, either consciously or subconsciously, and unimportant information must be disregarded or suppressed

- The signal should be able to access the relevant areas within the brain and appropriate action taken. This may be automatic or conscious, depending on the signal
- Short term, medium term and long term memory will need access and the mind will need to both retrieve and store information accurately
- Conversion from one sensory input to another is essential in some tasks e.g. the conversion of visual processing of a word to the "hearing" of the word in the mind, during reading.
- Feedback loops may be generated to control input sensitivity
- Attention and suppression must take place, instantly
- The mind must have sufficient information to take action as necessary - it may be crucial to survival!

The most important element in the above is that the input signal MUST be optimal, if it is corrupt, then every other element must be questioned. As we will see later, this signal is corrupt for the majority of those with reading problems.

Three dimensional processing

When looking directly at a picture of a face it is necessary to be able to perceive the relationship in space between given areas of the picture i.e. the right eye appears to be higher than the nose, to the right of the other eye and below the eyebrow. It is therefore necessary to have an ability to relate position within the visual system. In simple terms a map is required. From a processing perspective the sequence of the signals from the picture have to be organized sequentially at a neurological level. This sequential processing is mirrored from the ganglion cells at retinal level to the striate cortex in the brain. However, we can change the image into a moving image by "flashing" a consecutive series of images instead of an individual image. In other words the sequential processing works in two directions, one the sequence of position, the other the sequence of order. To complicate matters there is also a feedback loop from the brain to the eyes although the effect of this is not clear. I will postulate on the feedback mechanisms later.

The system can be considered to be a transient system (magnocellular) and sustained (parvocellular) in which the parvocellular is the central sharp vision that allows word discrimination, and the magnocellular that provides edge and movement discrimination, thereby allowing accurate eye coordination.

The parvo system is more sensitive to low frequency flicker. This facilitates fixation on stationary objects. The transient system responds better to high frequencies, that is fast flickering rates. This is useful in determining peripheral movement. Research suggests that dyslexics may have an abnormal flicker response in the transient system.

The sustained system is more sensitive to high spatial frequencies, whereas the transient system is more sensitive to low. This means that, in sustained vision, we are able to see greater detail and that ability to discriminate is enhanced.

The transient system has a large receptive field relative to those in the sustained system. These fields overlap as a three-dimensional mosaic. Fusion of these fields is essential both in each individual eye and as a binocular function. There is evidence that, in some, this fusion is aberrant. In the Jordan Reversal and Inversion test, monocular fusional anomalies are shown. This test also will demonstrate a phenomenon similar to the Troxler effect, in which the peripheral image changes from white to black!

The type of visual stimulus is critical, as the sustained system responds throughout the stimulus presentation. That is to say, it will respond all the time, while it can see an object. This contrasts with the transient system only responding at stimulus onset or cessation. This enables movement to be seen, very quickly, in peripheral vision. This important evolutionary effect enables potential prey or predators to be seen more easily. It is likely that each system has inhibitory affects on the other. The transmission of the neural impulse to the brain appears to be slower in the sustained system than in the transient system. There is a time delay in the signal arriving from the parvocellular probably due to the greater number of nerve endings in the nerve pathways.

We have shown experimentally that transient and sustained vision may have the potential to be separated in specific situations. This is particularly important, when dealing with computer operators or under fluorescent lights in a classroom or office. If flickering light, coupled with convergence and movement, can create a separation in the normally fused systems, it is likely that visual discomfort will be present and other symptoms may be provoked. Flickering lights can start migraines, headaches and epileptic seizures.

Patterns of high contrast will create a similar effect. The patterns that cause these difficulties produce predictable visual distortions in variable degrees of intensity, depending on the lighting and pattern design.

Recreation of symptoms

We decided to make an artificial "retina" using photographic plates to assess the effect of desynchronising the on/off areas experimentally. The effect was startling. All the distortions and visual effects of the patterns that are experienced in pattern glare were mimicked. There are many scientific problems with the procedure followed, but the results were so similar to the effects experienced, that I believe that further research is appropriate. I would be happy to pass on to any interested party the techniques we used. Retinal desynchronisation is important in many ways and the associated effects all seem to be related to visual dyslexia.

Image position

When you look at an image it is unlikely to be either static or positioned, directly in front of the eyes. The distance it is away from the eyes may be variable. This creates many difficulties with imaging. These include:-

- Differing image size relationships depending on position of gaze
- Differing focusing between the images resulting in poor image focus in one eye
- Differing alignment of peripheral objects on the retina depending on position of gaze and convergence, resulting in misaligned images
- Differing rotational needs of the eyes depending on convergence and position to maintain fixation
- Suppressional difficulties resulting in changes from pursuit to saccadal eye movement during convergence
- Difficulties in developing fixation resulting in muscle balance anomalies
- A "time lag" of loss of fixation during eye movement resulting in disappearance of objects, such as small balls

Controversially some squints, some constriction of the visual fields and some forms of nystagmus may also be consequential to difficulties of three dimensional processing.

In my opinion, it is also likely that magnocellular difficulties may ultimately prove to be a three dimensional processing problem. If this is the case, it would turn traditional magnocellular theories, that suggest, that visual dyslexia results from magnocellular defects that cause eyes to "slip", into an effect and not a cause, and it would follow that the change in the size of the magno cells in the lateral geniculate nucleus is the result of inappropriate stimulation in those cells, that are genetically or physically susceptible to change. This cause and effect will be significant, later, in explaining some of the symptoms of dyslexia.

Accommodation and convergence

In looking at any object it is necessary to focus (like focusing a magnifier) and rotate the eyes, until the central vision of each eye is in alignment with the other. The closer the position is, the more disparate will become the alignment of the peripheral vision between the eyes. This results in either suppression within the system or double vision.

When converging on a point, with a distant object still visible, there are significance changes in size of the distant image. This image may also be double. These images are colour dependent and may play a part in the ability for significant changes in convergence, seen by the use of colour (in particular the Optimeyes lamp).

Those with dyslexia often show a deficit in fusional reserves; i.e. they have less ability to deal with disparate images. Common indicators of visual dyslexia include

- Convergence reduction
- Exophoria (a muscle imbalance in which the eyes diverge)
- Strabismus and amblyopia (squint and lazy eye)
- Fixation difficulties
- Tracking problems

There are also some related eye positional difficulties that may be present. These include cyclophoria (rare) although head tilt may be indicative and may indicate a vestibular reflex anomaly. This responds well to treatment and immediate change in posture is the result. Nystagmoid (jerky) eye movement will sometimes respond well, although results to date have been idiosyncratic.

In general, I would advise any person with an eye movement anomaly to be assessed, because it is unlikely to be detrimental and results I have seen on patients that have "intractable" difficulties are sometimes astonishing.

Neurological difficulties

There are other neurological problems that may cause difficulties with visual processing. These include excessive blinking (blepharospasm), that will often respond to perceptual treatment and involuntary movements that sometimes can be helped. These are often found in Parkinsons disease and Dystonia. Visual field defects may also cause problems with perception, although some types will respond to colour treatment.

Eikonometrics and image formation

For many with problems, there is a significant difference between the two images in both horizontal and vertical planes and the position of gaze affects the plane skew. In other words a square box would appear as a parallelogram or rhomboid depending on position of gaze. Inevitable consequences are difficulties with proprioception and clumsiness, eye tracking problems and fixation difficulties. Exercises may be prescribed, after the person can see space properly, but it is inadvisable to use these before spatial efficacy has been determined. These exercises try to improve eye and movement control at the cerebellum, creating automaticity. The exercises take about 2-6 months to work and colour can often achieve the same effect, immediately, and at a fraction of the cost!

Often there is a perceptual magnification effect shifting the mid line and forcing the eccentric viewing, so often seen in autistic spectrum disorders. This can cause unsteadiness of gait and walking eccentrically and may be treated, immediately, using the Orthoscopics methodology.

There can also be changes in suppression, depending on position of gaze, resulting in eye switching at the mid line or failure to use one eye's central vision. In this case, perceptual expertise is necessary to determine the best way forward.

Image size differences are also important in squint and lazy eye, but are difficult for the optical professional to analyse. The Orthoscopics T test, can, in the correct hands, measure plane skew and predict expected improvements.

Attention

In looking at anything, we process the important information and to some extent disregard that which is superfluous. This allows less processing power to be used by the brain and the possibility to have a greater field of secondary information available. The fixated area that is accurately seen may be considered to be the area of attention, whilst reading will cover one or two words only. This area is inevitably smaller in people with reading difficulties or fusional problems.
Interestingly, in each case if improvement is made to the condition then the area of attention increases.

Colour

Colour is an extremely powerful tool, in the right hands. Evidence is now overwhelming that treatment, using properly prescribed colour, can have major effects on some people.

So, what is colour?

Colour is perceived by the mind as a response to electromagnetic waves in the visible spectrum. This means a wavelength ranging from 400nm to 700nm (There are one billion nanometers – nm – in one meter). There are three types of cone cell that respond to colour, one that responds most in the red region of the spectrum, one that responds most in the yellow/green region and one in the blue area. Every colour that can be seen is due to the responses of these cells.

In fact colour vision is extremely complex and beyond the scope of this book, but responses to colour are critical in the understanding and treatment of visual dyslexia.

We need to address how we define colour. Colour has shade(hue), brightness and saturation (density of shade).

It can be additive or subtractive i.e. can be created by adding colours together or filtering colour from an illuminant using tints.

When we define colour filters we need to be precise as to the colour mix at retinal level, as the effect of tint or overlay will change under different illumination (metamerism). This has major effects on assessment and treatment and, at present, the only method of assessment that takes this on board is the Orthoscopics method (see later). Metamerism also makes it impossible to use tinted spectacles or overlays in an accurate assessment and explains why colour of tinted lenses rarely match overlays. I would suggest overlays are ONLY acceptable as a short term treatment.

Colour can also be important in focusing and convergence (chromostereopsis) and colour relationships can provide information as to position. Dispersion of colour (chromatic aberration) may have some influence on accommodation. Coloured text or paper may therefore affect, both adversely or positively, the ability of the visual system to work efficiently. The illumination of the text by background lighting is critical for some and, therefore, all assessments must be undertaken in optimum lighting conditions. These vary from one person to another. It follows that it is essential visual performance assessed immediately problems occur, and accurate determination of optimal colour conditions be ascertained.

Colour constancy (seeing colours as not changing under different illumination) means that testing must be performed in a dark room and ideally using additive colour (not by using filters).

Although there is no suggestion that visual dyslexia is a colour vision problem it may be related in many cases. In many people with visual dyslexia there is a significant difference in either brightness or colour perceived, with each individual eye, relative to the other. This effect is common and may well be related to retinal rivalry. When monocular treatment is suggested, invariably, it is more satisfactory to treat the eye in which the image is too bright. Frontal headaches are common when these symptoms are significantly strong. Research in this area is scant and not conclusive.

Suppression

It is necessary to see the whole text, during reading, but what happens to the unwanted information? Different forms of suppression are necessary in processing information.

Earlier, we stated that vision is a learned function. Suppression can also be learned. Some suppressional effects can be harmful, others beneficial. When reading we have to suppress the unwanted information around the words, as, without this suppression, constant double vision will occur. We also have to suppress the imagery from the binocular vision, that occurs during reading, particularly compression.

Memory and working memory

When a person views an object or word, they have to process the information in two ways.

- Use their memory to make sense of information presented
- Put any information into the memory, from the information presented

He or she may have to access motor areas to respond with movement, speech and language areas, smell and taste areas or "thinking" areas.

All these areas can be affected by visual stimulus and modifying the stimulus can have surprising effects in rare cases e.g. we have seen symptoms consistent with Wernicke's aphasia (talking gobbledegook) being provoked by inappropriate visual stimulation.

Synesthesia (mixing up of neural pathways) can often change remarkably using visual stimulation. Memory and sequencing, such as digit-span, is more-often-than-not changeable, immediately, using modification of visual input. Visualization can also be improved with immediate effect.

Peripheral vision

In addition to central fusion being improved, there is evidence that significant changes to visual fields are possible by colour change. This is of particular importance in driving, as the standard optometric tests for visual field problems are inadequate, when colour field problems are present. It also partly explains the difficulties many have with ball sports.

Spatial characteristics

The ways, in which spatial characteristics of text and environment cause a change in perception, are predictable and important in the assessment and treatment of visual dyslexia. The higher the contrast between the text and background, the greater the risk of difficulties and, the larger the area covered, the greater the risk. The frequency of the pattern, the lighting and the person's susceptibility are all influential in determining the degree of problem. The environment is also a factor, with one of the worst possible environments being the average school classroom! Flicker is also a potential problem and can be introduced by eye movement on high contrast text. Therefore, a change in text presentation must be considered and the classroom should be optimised as an environment, for any child with problems. It is also clear that differing results can be produced by the same reading tests using differing fonts

Reading

A book is usually held at between 20 cm and 40 cm from the eyes, depending on height and therefore the length of the arms of the reader. To be able to read, the eye has to be able to have a clear image on the retina and consequently has to change the vergences (the cone angle) of the light, by the process of accommodation. Steepening the curvature of the lens within the eye changes focus to a closer point. This ability to change the shape of the lens within the eye reduces with age, from the ability of a baby to focus at 5 cm to that of a pensioner who only can focus at one metre. The degree of focusing ability, will actually allow age prediction.

This focusing ability is also influenced by pupil size (the smaller the pupil, the greater the depth of focus). Pupil size is constantly changing, becoming smaller during accommodation. At the same time the eyes rotate to a point in which both optical axes look at the same position in a word to allow foveal (that is central sustained parvo vision) to take place. This point is called the fixation point.

The ability to maintain fixation on this point may be erratic in young children and those with visual dyslexia. This difficulty can be helped in the majority of cases, relatively easily, by using visual processing modification. During the rotation to the fixation point it is necessary to stabilize the peripheral vision by suppression of peripheral vision in one eye. This is a learnt action. Alternatively fixation may be present in the central vision of one eye only (in conjunction with peripheral suppression, if necessary, to achieve stable vision). If this learnt process does not achieve stable vision or is incomplete, it is inevitable that unstable fixation will be present. The predicted symptoms of this form of unstable vision are identical to those shown by some children with reading difficulties.

The effects in reading of Anisometropia and Anisokonia.

Anisometropia is where there is a large difference in refractive error between the eyes. If the eyes are not corrected, then amblyopia will be the conclusion, in many cases. However, correcting the eyes may, in some cases, create a situation in which the symptoms of visual dyslexia are produced. This is due to the inability to fuse the unequal image sizes produced by the different magnifications in the spectacle lenses.

As the eyes move from the central position, prismatic effects create a relative positional change of objects. Shapes are distorted and, as the eyes move, the distortion reverses. Fusion of the images is difficult in certain positions, particularly if the disparity in refraction is vertical. Contact lenses will often resolve the problem.

In older people this can be an even greater difficulty, than with children, as they are often prescribed bifocals or varifocals. As their eyes are depressed to see through the reading area of the lenses the images may separate and give double or unstable vision or, alternatively, suppression of one eye may result. It is sad to say that very few get optimum treatment of this problem. A rare problem is that of a person with a similar refractive error in each eye who has a marked difference in image size. In these cases visual stability is impossible, fixation is impossible and constant distortion is present. Specialized visual treatment can resolve many of the symptoms and occlusion or colour are often appropriate.

Anisometropia and anisokonia have similar visual processing effects to visual dyslexia suggesting that similar fusional problems may be exist.

Dynamic vision

During saccadal vision in a block of text there are a number of processes, other than eye movement. The vertical lines in the print will create a stroboscopic effect, as they stimulate the magnocellular system with their hard black and white edges. The peripheral retina is extremely sensitive to flicker and has to be able to be suppressed during the eye movement in reading. The immediate area of the fixated word is usually in the central visual area; the larger this area of central vision, the easier it is to read. The size of this area is determined by fusional abilities as well as the anatomical structure of the eye. The better the fusion, the larger the area. Visual processing improvements often immediately increase the size of this area.

The gap words between words clearly fixated are interpreted by the brain by virtue of their shape. Reverse saccades "fill in the gaps" of words.

Visual stabilization is necessary and ocular dominance will often develop as a learned response. Right eye dominance is the most common and will result in fewer visual processing problems than left eye dominance. Occasionally, dominance does not develop, with some resulting difficulties. A simple, although not foolproof, test for dominance is to check which eye a child will use to shoot an imaginary gun.

In addition there is a relative accommodative change of a quarter to a half dioptre between the eyes as they traverse the mid-line. Image size differences can be in the region of fifty percent between the eyes causing fusional difficulties. Lag of focusing can also be present.

Retinal rivalry

Retinal rivalry will sometimes create an anomalous effect by projecting the image from one eye into the perception of the other. This can be utilized in treatments such as monocular partial occlusion. Unfortunately the mechanisms of retinal rivalry have had little research. Reversed retinal wiring may influence interpretation of letter shapes. This may also apply to sequential interpretation although this is more likely to be a memory or neurological problem.

In tracking along the line, many children miss words out or give incorrect ends of words. They may find that the letters appear to be more compressed towards the ends of words. Suggestions have been made that it is a magno-cellular problem, but this is unlikely.

.

The relationship between input stimulus and performance.

Some suppression may be due to a hypersensitive processing system. There is a relationship between input information and response. As we increase the input, the response increases to a peak, which is followed by a drop. This reduction in response levels means that the entire input stimulus must be targeted to achieve maximum response. Therefore, too much stimulation is as detrimental to performance as too little. The critical thresholds for optimum performance vary depending on the person and the type of stimulus. The most appropriate lighting conditions must be individually determined and the critical threshold for hypersensitivity may be achieved much earlier in some individuals than to others. It is likely that many people with visual forms of dyslexia are hypersensitive, although some may be hyposensitive.

There is a continuum between stimulus input and effect and, therefore, it is possible to recreate the symptoms of visual overstimulation in anyone. We all have the potential to be visually dyslexic!

Extreme Symptoms

Certainly, there can be differences in symptoms exhibited. Extreme visual processing problems are common e.g. one unfortunate patient of mine suppressed neither eye but had the misfortune to have total inversion of the print at reading, only with one eye. The other eye maintained the print, correctly orientated. Consequently, she had significant difficulties reading. Occlusion of the eye, that saw the text upside down, ensured that she could read perfectly well and her problem was effectively resolved. This inversion did not occur in distance vision and only occurred on convergence. We have found inversion in many people and it is likely that it is much more common than we had realized

Another example of an extreme problem was that of a young man, who had just left school, having grossly underachieved. He has an IQ that would normally get him into university. At sixteen, he could read flashcards but not text. Medication was used as a chemical cosh because of his behavioural problems. Suicide was attempted. His description of the printed text was "it is like watching a washing machine spinning, it's faster with the right eye". Within twenty minutes he was reading text as his vision was stabilized. It still upsets me to think of what that boy must have been through. It is vitally important that we stop children suffering unnecessarily, we owe it to them.

Cognition

When we have the optimum visual input it is necessary to achieve understanding of the information presented. In an ideal world this would be perfect, after we have modified the visual input. This understanding and comprehension is often flawed and this is the point, at which other professionals should take over from the optical professional. Educational and psychological techniques are beyond the scope of this book.

Chapter 4

The senses of a hunter

History of visual/auditory combined pathways.

The development of the human visual and auditory systems were governed by his need to eat, shelter and produce healthy offspring.

His visual demands were likely to be gender-linked with the male being more likely to be the hunter and the female the gatherer and child rearer. In an evolutionary context, if we consider the visual needs of a hunter or gatherer they obviously are not the same as a modern office worker.

There is one crucial sex linked difference; gatherers (girls) spend a lot of their life with their visual system working within arm length, the hunters (boys) visual system is used for distance viewing. Is this why the visual systems have developed in such a way as to make boys four times as likely as girls to show symptoms of visual dyslexia?

The visual requirements of a hunter

To be a successful carnivore, the hunter has to be able to defeat his prey mentally and physically. First, he has to locate his prey. In man, the visual system is sensitive to movement in his peripheral visual fields. If movement is detected, the head and eyes rotate to fix the movement in the central visual areas. Peripheral hearing fields and visual fields overlap to determine prey or predator position i.e. some cells that respond to visual movement, also respond to noise, thereby allowing fast and accurate positioning of the potential adversary.

The central areas are incredibly sensitive to form, allowing shape recognition to determine whether the movement is from prey, a predator or not important to the person. Binocular vision and other visual clues produce depth perception. We can guess the range of our prey.

Smooth pursuit movement allows us to fixate and follow moving objects but there is little need for saccadal eye movement (jerky eye movement in reading). Accommodation is used to its maximum potential infrequently, as most important events are at a relatively long range. Convergence is rarely used to an extent that extends the individual's convergence reserves.

Compare these visual tasks with the modern schoolchild. Virtually all critical information in schoolwork is presented to the child in their central visual areas. Convergence and accommodation are stretched to their limits, continuously, saccadal movement is utilized, virtually all the time, instead of pursuit movement. The lights and computer screens flicker at a rate that is disturbing and the text creates patterns, that are often a direct cause of visual stress. It is surprising, that so few have major problems. It is a completely different environment from that to which our visual system is optimally tuned.

Auditory processing.

The hunter also relies on his auditory skills. Generally, during the hunt, he will be extremely quiet, as this will allow him to hear noise produced by his prey and will keep his position from being compromised. The relative volumes from each ear will allow him to assess position by rotating his head to equalize the noise from either ear, thereby allowing visual fixation on the point from which he believes the noise has originated. The subconscious timing difference of the signal from each ear enables spatial position to be determined. This awareness must have been critical for the survival of many of our ancestors. Again, this contrasts with the noisy modern environment in which these skills are obsolete and it is necessary to inhibit our perception of background noise.

Speech perception requires a focus on the speaker and a suppression of the background information in the auditory system. Reading requires precisely the same ability to read and perceive the word in the body of text, without the background information overwhelming the visual system. The parallels are obvious. The act of reading involves the same type of processing as that of listening in a noisy environment.

Is reading a visual or auditory activity?

When we read a word we have to undertake a number of visual processing activities. So, how do we read?
 The first visual action is to fixate on a word. This central gaze focuses on the first letter or the centre of a word. The

background information i.e. the text around the word has to be ignored although context will become critical as reading speed increases and words adjacent to the fixated words become more important.

All words may be recognized by one of two ways, they can be broken down into component sounds and recombined in the mind (phonics), or, the shape of the word may be recognized if it is common (shape or character recognition).

The word is then "heard" in the mind. In other words the mind transfers the visual recognition into the auditory processing system and thereby into the auditory memory. This contrasts with using the visual memory of an event; which is entered into the visual memory. Therefore, although reading is initially a visual activity, we find it necessary to change the visual stimulus into an auditory internal sound to enable words to be recognized. When we read, we don't actually see the word in our visual memory, we actually hear the word in our mind. In children with visual dyslexia, it is common that they use the visual memory rather than the auditory memory.

Tracking

Words appear bigger on the same side of the page as the eye i.e. the words on the right side of the page look larger to the right eye. This means that words on the right side of a word or the page appear smaller to the left eye and may be a reason why tracking is more difficult for a child with a dominant left eye. The accommodative effort varies substantially between the eyes as a trigonometric function as the eyes move along a line of text.

There is a change in the hardness of the edges that affects Magnocellular processing. Reading places severe demands on the visual processing system.

Memory

Short-term memory problems are common in all types of dyslexia; indeed, some believe that dyslexia is principally a memory problem. Sounds that we "hear", through our eyes converting our visual world into a hearing world, change our visual memory to an auditory memory, representing an idea, a concept, or a shape. In primitive activities, such as hunting, there is little use of the transfer of visual stimuli to the auditory memory.

We can however use these primitive memory switches to our advantage. In a simple case we can transfer auditory memory into the visual memory by the use of visualization. e.g. in a case of poor retention of sequential numbers in auditory short-term memory. The spoken numbers are converted to a picture, by closing our eyes and visualizing a projected image in our mind's eye. Using our hands, to pretend to draw the picture of the numbers or letters, re-enforces the visual memory. The person may then access the visual memory if required and can often achieve better results than using only auditory memory.

Synesthesia

A few people convert auditory input or characters to visual stimuli. This condition, known as synesthesia, involves people hearing tones or sounds as colours or perceiving numbers or letters as colours. In other words they can change the form of stimulus from auditory to visual or from form to colour, a reverse of the reading transfer of stimulus. It is, perhaps, more common than is generally reported, as most people, who have these effects, do not admit to them, as ridicule would be a likely result. (I found my wife and one of my children see numbers as colours. I did not know that they had experienced this phenomenon, until a discussion during the writing of this book!). Synesthesia may also allow some to smell in colour!

Development of visual dyslexia

If infant's eyes do not receive light and pattern stimulus, the visual pathways do not develop and they become blind. This form of blindness is called amblyopia (although amblyopia can be also used in other cases of visual acuity drop, without blindness).

Light, flicker and form stimulation are necessary, to ensure visual function developing normally in an eye. Therefore it can be said that vision and visual function is developed as a response to stimulation. There is a critical need for this stimulation to be undertaken, as early as possible, because the ability of the visual system to develop (plasticity) reduces quickly.

The plasticity of visual development is greatest in an infant and conventional theory states that all major improvements are complete, by ten years of age.(This statement is not true for all individuals). In other words, if you do not provide adequate stimulation it will be impossible for your visual system to develop properly e.g. in congenital cataract the light is prevented from reaching the eye and deep amblyopia, resulting in blindness, is the consequence.

There are a number of other causes of amblyopia

- Relative magnification differences between the eyes being too great to fuse
- Misalignment of the optical axes of the eyes
- Colour fusional areas desynchronised (not provenand may be unilateral -one eyed- or bilateral - both eyed)

It may be said that amblyopia can be caused by too much, too little or inappropriate stimulation.

Treatments include:

- Spectacles in the case of refractive error

- Occlusion to try and force the amblyopic eye to work. Incorrectly used, this could be considered to be child abuse.

- Surgery, to align optical axes (muscles around the eye are cut or their function is reduced to enable the eyes to line up their visual axes). This is used when a strabismus (squint) is present.

- Professionally prescribed colour can often improve visual acuity, immediately, and may straighten some squints (a VERY controversial statement, but I have shown this happening at a number of optometric conferences).

- Intermittent photopic stimulation, which provides a high level of stimulus to an eye to try and force the vision to improve.

- Contact lenses to equalize image sizes and stop differential prismatic effects

The first three are common techniques; the last three are used much less, although they often give superb and far superior results.

Why is this?

The last three are difficult for the clinician to administer. They are time consuming and do not always work. However, they are so infrequently prescribed, that it appears that many do not consider these simple alternative treatments.

It is clear that many treatments for strabismus or amblyopia in young children may actually be doomed to failure, because it would be impossible to create the optimum stimulus in which the visual system could develop correctly. For example, a child with poor vision in one eye is often treated by occluding or patching his good eye, for long periods of time, in an attempt to try and force the eye, which doesn't see properly, into working. Success is idiosyncratic. If the good eye is occluded, you may reduce that child's ability to see the world dramatically. This can be abusing the child, if prospects of satisfactory results are poor. I have seen children walking round the streets, effectively blind, in an attempt to improve their poor eye. The parents are often unaware of the effects on the child or knowledgeable about the cost / benefit of the treatment

On a personal note, I would not use occlusion or accept surgery on my child, without the consideration of alternative treatments. It is necessary to look at the causes of the lazy eye as similar to the causes of visual dyslexia and, therefore, an attempt to correct a lazy eye will have direct parallels in the treatment of visual dyslexia and will often be successful. Conversely, treatment for visual dyslexia will often improve the vision in a lazy eye.

Amblyopia is closely related to visual dyslexia Indeed, they may be considered interlinked in a high proportion of cases. This means that many visual conditions require careful analysis of the stimulus, to achieve optimum results.

The lazy ear

It is likely that a similar process to that of amblyopia happens within the auditory system. If a person can't hear a sound, they can't say it, if they can't say it, they can't read it and perceive it. The auditory system requires stimulation. These needs for stimulus are particularly important in young children. When a child experiences an infective ear problem, surgical treatment is not prescribed in many cases, as often the condition is self-limiting and will resolve spontaneously.

There are potential benefits in early treatment, as it is likely that the hearing of the child will improve and, as a consequence, his auditory processing will be better. Without resolution of the early auditory difficulties, many of these children will have auditory processing problems, which may lead to speech and language difficulties and, consequently, they are much more prone to dyslexia caused by the inability to hear correctly. Is non-intervention the most appropriate medical response? This dilemma is irresolvable.

Treatment

The next question is, at what age is it desirable to treat a lazy eye or a lazy ear? Standard treatments usually stop at around nine years old. Certainly, it is possible to improve lazy eyes well beyond the time in which the present treatments are discontinued. Dramatic improvements can occur, in some cases, by using stimulation and modification of stimulation techniques. Dynamic theory, that will be explained later in this book, may hold some benefits, beyond dyslexia and the visual aspects of dyslexia, in the future.

Allergies

A high proportion of sufferers from dyslexia exhibit allergic responses and they are often atopic, which means they often suffer from asthma, eczema and hay-fever. It is prudent to try and reduce the symptoms of allergies, with any child with visual problems. Although the link has not been confirmed, it is likely that there is some common ground. Certainly modification of neural transmission in allergic responses, by use of antihistamines (in high doses), has been used as a treatment in dyslexia. Results have been idiosyncratic and, to some extent, the jury is out as to their efficacy.

Substances, that mimic neural transmitters, often have effects on children with visual dyslexia. There are some children, who find that visual dyslexia is much worse, if they ingest significant levels of caffeine or food additives. Particular emphasis should be placed on the avoidance of tartrazene and red, yellow and orange food dyes.

Due to the effect of neural transmission problems in dyslexia, it has been suggested that using fatty acids as a dietary supplement may often help. These fatty acids are essential in nerve development and many dyslexics may be deficient. Supplements can have a significant effect in some but, again, results are erratic.

Neural pathways

Neural pathways in the brain develop through activity and may be lost through inactivity. It is critical, therefore, that as many pathways are established, as early as possible in life, as the body cannot increase or maintain this network, to the same extent, as it ages. The visual area of the brain occupies over half of the cortex and is, constantly, in a state of active communication. There are loops that form from the visual inputs to Wernicke's area (the area that deals with speech recognition) and Broca's area (the area that deals with language production).

There are areas of the brain responsible for each of the senses, although the neural network ensures that these areas are connected with other areas, in a mesh-like way.

The brain is divided into two sides with the division running through the mid line of the body. The two sides communicate through the corpus collosum. This busy highway is filled with cells passing "messages" from one side of the brain to the other (it is about fifty percent larger in girls than boys).

Although we talk about areas such as the visual cortex (the area responsible for vision), the nerve "messages" do not go just to one centre, they will go to multiple centres and will be coded in a fashion that takes into account the multiple inputs from the

82

neurons. Although we can see that there is right/ left communication across the brain, it is perhaps more complex than the simplistic view of looking at precisely one side of the brain only. However the structural differences in communication may have a bearing on dyslexia, particularly as girls have better communication across the centre line. Girls are less prone to visual dyslexia, but suffer many more migraines.

Frontal headaches and migraines

 In many people, frontal headaches are common, if they have visual problems associated with dyslexia. These occur in approximately the same area as the pain in sinusitis. It is interesting to note that, in a very high proportion of those with diagnosed sinusitis, there does not appear to be any clinical sign of sinusitis, in post mortem examination. Treatment for visual dyslexia resolves the vast majority of frontal headaches, including the headaches of many sufferers of sinusitis, and migraines often disappear completely, or reduce in frequency or intensity.

Are many migraines another form of visual dyslexia? It appears that a substantial proportion of these may be closely related, as they can be treated as a form of visual dyslexia and a resolution of symptoms is often immediate. After all, the classic treatment for migraine is to go into a dark quiet room and rest! In practice, I treat migraine as a form of visual dyslexia. The success rate is high and immediate, but not everyone benefits. It is however apparent that a high proportion of migraine sufferers have major problems, due to light and pattern, and, virtually invariably, these are helped by treatment.

The medical profession and migraine sufferers often resort to medication, too quickly! Specific filters and in particular the Optimeyes will often stop a migraine, without any side effects. Many opticians have known for years, that in some people the headaches will resolve immediately, if filters, exercises or optical prescriptions are used. However, the prescribing of tints was undertaken by trial and error techniques and, as a consequence, the responses of patients were erratic.

Later, we will try to give a rationale on how to look for the visual causes of migraine and how to eliminate the problems associated with visually induced migraine. We can also extend this relationship to people who from suffer travel sickness, agoraphobics and panic attack victims.

The trigeminal effect

The trigeminal nerve is a complex cranial nerve with three branches. The motor portion is mainly concerned with chewing, the sensory with proprioception (spatial knowledge), touch, pain and temperature from the eye and mouth areas. Trigeminal neuralgia (pain) is a known response. The anecdotal and circumstantial evidence, that some forms of visual dyslexia exhibit abnormal trigeminal nerve enervation, seems to be overwhelming. It may be that many of the symptoms of visual dyslexia are produced in association with or by the trigeminal nerve. In relation to visual dyslexia, what does the trigeminal nerve do?

It controls the tensor timpani muscle, responsible for the tonus of the ear drum. In patients treated using filters, their ability to hear tones can be modified.

It controls tear production in association with the facial nerve. Dry eye symptoms are treatable using filters, with immediate improvement for some. In children, the symptoms of visual dyslexia include hot itchy eyes that resolve during treatment. Red eyes are common in children with reading difficulties!

It is implicated in migraines/ frontal headaches. Trigeminal neuralgia is a common painful face-ache, which may be on the spectrum of asthenopia and migraine.
There are a number of relationships with other sensory nerve pathways, that appear to be affected by the trigeminal and visual innervation nerve. These include effects on balance and hearing.

Vestibular system

Some vestibular processing problems can create balance difficulties. Visual treatment e.g. prisms or filters, that affect the convergence reflex, will often help because of the inter-relationship between the eye and the ear in the vestibular system.

Positional sense of the head is determined by the vestibular and proprioceptive systems. The balance system integrates with the visual input and the proprioceptive system, to enable positional stability. The eyes rotate as a response to head inclination. It is likely that perception also plays a part in determining knowledge of head position. The perception of horizontal and vertical surfaces will allow the mind to assess orientation. If there is a problem with the semi circular canal input, visual perception will often override the anomalous input, but, if the eye does not fully compensate, then balance problems will still persist.

Modification of visual stimulus will often modify the response, with an immediate effect. Balance can be grossly affected if the visual processing system is inadequate and may be treated, in some cases, as a visual anomaly.

It is essential to realize that auditory processing is effected by visual input.

All opticians have heard the statement" I cannot hear properly without my glasses" The general belief amongst opticians is that the person lip-reads. This is probably untrue in most cases. It is certain that the convergence reflex affects tonal discrimination, in some people. e.g. If we assess a person with convergence difficulties, it likely that that person will also have problems with hearing and they will perceive tonal differences poorly.

Optometric signs

At an early age, it is extremely difficult to determine whether visual dyslexia is present and it is generally only the most obvious cases that will be detected. However, as a number of visual and other conditions are related and these are more easily detected, it is necessary to take a more rounded approach to early diagnosis and intervention.
These conditions include strabismus and amblyopia.

Strabismus (squint) is an inability of the eyes to maintain alignment of the central vision. There are many types of strabismus, the most common being convergent, divergent and alternating.

Convergent strabismus. This is due to hyperopic refractive error (long sightedness) and can often be corrected by spectacles. The eyes rotate towards the nose and the brain suppresses the central vision from one eye. The peripheral vision from that eye is still used for movement recognition. Should alignment not be achieved, at an early age, the central vision will be deeply suppressed and poor vision in the squinting eye will become permanent. If the refractive error is significantly greater in one eye than the other, treatment with spectacles often gives poor results, due to an excessive difference in the relative magnification of the spectacle lenses. Spectacle lenses will also produce a prismatic effect, due to the shape of the lenses, which will often be difficult or impossible for the brain to fuse. In patients with a difference of four dioptres between their eyes the functional area fuseable at a reading position is only about 2.5mm by 5mm. It is no wonder that the success rate is idiosyncratic. Contact lenses have a much greater chance of success but are relatively rarely fitted due to the cost and great skill required in fitting. Convergent squint sufferers often exhibit many of the symptoms associated with visual dyslexia. Colour will sometimes improve results, significantly, and will often stop the visual dyslexia symptoms that are so often found in those with convergent strabismus.

In divergent squint the eyes will separate with one eye fixating and the other looking outwards. The diverging eye will suppress the central vision although the peripheral vision will still be used in that eye. It may be caused by inability to see or fixate with one eye, although uncorrected myopia (short sight) may be responsible. Anecdotal evidence seems to suggest that visual dyslexia is less relevant in this form of squint but they do occasionally coincide.

Alternating strabismus. This is where the brain alternates in suppression and non -suppression of the visual signal coming from each of the eyes. This form of squint may sometimes be a reverse of the effects of visual dyslexia, for distant viewing, as the person will sometimes have stable binocular vision in the near field.

It can be argued that, if both eyes are stable both peripherally and centrally during reading, then it is impossible for them to be stable in distant viewing, without suppression. This is a reversal of the normal binocular stabilization in the distance and suppression of either central or peripheral images for near vision and appears to be a potential causative factor in alternating squint. Those with alternating strabismus will often show visual perceptual symptoms that can be corrected with properly prescribed colour.

It is unlikely that your child will be offered colour therapy for binocular vision problems, as the science is in its infancy in this area, but I personally would insist that it be considered. Many in the hospital system will be unaware of how to use colour efficiently – ask your professional if they have been trained in colour therapy to ensure that they are aware of this new area of expertise. I could only recommend the Orthoscopics system for an assessment and associated colour therapy.

Interestingly, we have identified at least two methods of suppression when reading text. These coincide with theoretical expectations. In standard optometric tests I do not suppress central vision, when reading, but, when confronted by a block of text, I suppress the central vision of my non-dominant eye. My peripheral vision is stabilized and I do not have any symptoms of pattern glare. This can be contrasted with an optometric colleague's suppression for near vision. Central vision was stabilized in both eyes, with suppression of one of the peripheral retinal signals. This produced a reduction in the size of the central visual field, instability of peripheral fields and significant levels of pattern glare. However, using standard optometric tests, our vision was effectively identical. This demonstrates one of the inadequacies of optometric examinations: these are not a visual processing assessment.

An eye examination, that results in a statement that there is no disease or abnormality, does not mean that visual processing is performing optimally. An eye examination does not assess visual perception and it is vitally important that the optometrist refer the child, with appropriate symptoms, to a person that can assess, if necessary. It is common to see children and to arrive at the erroneous impression, that there is no visual problem, when there is a significant unresolved processing problem. This is particularly so, as the parent is not told of the differences between pathology, physiology and processing. The optometrist has a duty to inform the patient of the limitations of a standard eye examination and be aware of how to proceed if a processing problem is suspected.

The parent of a child with reading problems must insist on a full processing assessment as well as a full eye examination, although it must be remembered that further costs will often be incurred and that these may be considerable. An eye examination would normally take twenty to twenty five minutes; a processing assessment would be as least as long again.

What symptoms should make us suspicious?

Many symptoms of visual dyslexia are also present in eye or systemic disease. There is an absolute necessity for an eye examination to rule out pathology, before considering processing problems. It is also necessary to have the ability to recognize potential signs and symptoms of disease, during the processing assessment. This requires a high degree of knowledge, although screening is straightforward.

Symptoms of visual dyslexia develop both as the child matures and as the visual demands become greater.

Chapter 5

Development

Visual dyslexia gives differing symptoms, depending on age

A significant problem with light (photophobia) may be suspicious in an infant. There are other more likely causes of these symptoms e.g. infection.

A young child may rub his eyes and complain of headaches in the temple areas. This is often found in conjunction with an abnormally large pupil.

Allergies or sensitivities may predispose a child to later problems.

Hyperactivity and attention deficit are highly suspicious; diet must be monitored rigorously in these cases, as many problems can be resolved by eliminating the offending food. If foodstuffs give symptoms of a sore stomach or temporal headache, reduction or cessation of intake is indicated. The foods that appear to create the greatest problems are:

- Milk or milk products
- Gluten

- Citrus fruit
- Food colourings - particularly orange, yellow and red
- Caffeine
- Cheese (and red wine)
- Onions and garlic
- Chocolate
- Cheese

As a child enters school, visual processing becomes more demanding. The environment provokes visual processing difficulties. As well as the emotional upheaval, there is the physical insult that is often experienced, which is caused by the design of the classroom. The types of lighting, with low frequency fluorescent lamps, are particularly problematical for some. Many children see these lights flickering, constantly, and live in a stroboscopic classroom. Health and safety issues are obvious consequences. The child has a right to a safe environment.

Teaching techniques and materials e.g. white boards, paper type, font style and size, can create the worst possible re-enforcement of the environmental injury. It is little wonder, that for some, school life is a miserable time. I believe that, in general, the teaching profession does not realize the unpleasant physical time their pupils undergo and will be horrified, when they realize that their classroom and their teaching materials and techniques contribute to the real problems that these children endure.

The classroom problem

The classroom lighting levels, required in the UK, are 300 to 500 lux. This is generally achieved by the use of standard fluorescent tubes. These have an uneven emission spectrum that is characterized by spikes (large increases in emission at very specific wavelengths). The tubes flicker at 100 Hz i.e. they switch on and off one hundred times a second, assuming that they are properly maintained. Poor maintenance will often result in lower frequency oscillation. It has now been shown that some people are sensitive to this flicker. The result of this sensitivity varies from epilepsy through migraine, headache and visual discomfort. It has also been shown, that some people are sensitive to particular wavelengths of light and, where these are coincident with the emission peaks, visual stress is inevitable.

Some fluorescent tubes also produce a constant noise that may cause extreme discomfort in some susceptible people. Poor maintenance may be contributory. For the biggest improvement possible in school - rip out these lights – high frequency alternatives are available.

The architectural design of modern buildings may also play a part in producing an unpleasant visual processing environment. Patterns that produce stress are often unwittingly incorporated into the sleek interiors of modern buildings and may be, in part, responsible for "sick building" syndrome. Certainly, it is possible to reduce workplace migraines, headaches and discomfort, if modification of the visual input is undertaken. Perhaps it would be better if architects were trained in the effects of their designs on the visual system as some of their designs inevitably cause visual stress, headaches and migraines.

Computer screens that are refreshed at a low frequency, or have an unstable flicker, may add to this stressful environment. Computer justified texts are more likely to create difficulties and high quality "bright" papers increase the likelihood of glare-producing text, due to the increase in contrast between the print and the paper. Giving a child text printed on coloured paper can help, but every child needs to have the colour individually determined. One colour is not acceptable for everyone although an "off white" paper and greyer print would be of assistance, as the contrast is reduced. Increasing the size of the print is of vital importance; many symptoms of visual dyslexia can be stopped by magnification.

The noise levels in the classroom contribute to the problems experienced. For many children it is impossible to filter out the background "clutter". This results in an overload of the auditory processing system and reduces performance. In conjunction with the visual overload, that is also often produced, many will find the classroom an extremely unpleasant place. It is little wonder that these children often become disruptive or withdrawn. After all, they have had an assault on their processing systems. There can be no excuse for not creating a reasonable environment for our children and consequently for our teachers. If we get it right for the children, then the working environment will be improved for the teachers and the results will be fewer migraines and headaches and consequently a happier, less stressed and healthier teacher.

Early reading.

As a child develops, his short-term memory also develops. A general guide is that, up to seven years of age, a child will recall one extra digit or character in a string per year (plus or minus two). In other words by the age of seven it should be possible recall between five and nine characters. Sorting of these characters, e.g. putting letters in alphabetical order, will usually reduce one digit from recall ability. This ability will normally be equal between visual and auditory processing. If the child has significantly reduced recall ability, it may be impossible to read phonetically. Teachers must be aware of the importance of recall and the techniques that can be used to enhance memory.

Transfer of visual information into the auditory system may be poor in some, who have visual or auditory processing problems. Reading out loud for longer often will help this transfer of information, as the pathways have to be established between the two processing systems. Some children do not hear the words that are seen easily. This skill is essential for good reading.

There are problems related to poor visual tracking. This manifests itself through difficulties in moving along a line of text and, sometimes, jumping to another line. As the eye traverses the line left to right, the image in the left eye first gets bigger, reaches a maximum size in front of the left eye and gradually gets smaller to the right side. This is reversed in the right eye. Fusion is difficult and, depending on dominance and fusional ability, the eyes may slip or jump.

The other effect, due to this phenomenon, is the difficulty in seeing the ends of words. The child will guess the ends of words such as the, that, those for their. These symptoms may also be caused by recall memory difficulties although it seems strange that the effect seem limited to words of similar visual shapes. This may be why words are missed out in reading, although, if a visual saccade (jump) is too large, it may prove impossible to perceive all the words between fixation points. It may also be due to neurological organisational problems.

In early reading many children will reverse letters or words. In recent times, this has been considered as a memory problem. This is incorrect.

A new test, the Jordan reversal and inversion test demonstrates, conclusively, that some reversals are due to wiring anomalies in the visual system. This test is described later in the book.

 The reversals and inversions of letters e.g. d/b p/q t/f are treatable, in seconds, by stimulus modification e.g. magnification, minification, colour modification, changing lights. At present, we find that about twenty five percent of those with reading difficulties show reversals, either in one or both eyes.

The second form of reversal is due to binocular suppressional difficulties. In certain conditions of crossing the mid line, when reading, one eye sees a reversal of letter order relative to the other. This only applies to short words e.g. was/saw. If visual suppression alternates, the brain will interpret the word incorrectly; although it is more likely that, in some cases, these types of reversal are retrieval anomalies from the lexicon (short common words can be recognized due to their shape without using phonics).

The third form of reversal is sequential memory reversal. This is due to the incorrect input or retrieval of the order of the letters in a word. Compression and expansion may be responsible for incorrect input, although neural network wiring anomalies appear to be a more likely cause of the problem. These symptoms must not be ignored.

There may also be another explanation for sequential letter problems in some cases. Recently we have found, that modification of the light source for some people changes the sequential visual order. This change takes place spectacularly, in front of their eyes!

The early reader with these visual difficulties, then, has to contend with the consequences. He will obviously have difficulties in concentration, his self-esteem is lowered and he may become disruptive. The cycle of underachievement has begun!

The teacher will often criticize, not recognizing that the child has a disability that causes a major problem. This will usually continue, throughout the child's school life, as it is rare for all but the worst to be recognized as having this problem. I suspect that less than five percent of those with significant visual perceptual difficulties get any help. This is totally unacceptable and must be changed.

As children get older, less profound sufferers start experiencing difficulties. Texts become smaller and in larger blocks. This creates the perfect environment for pattern glare symptoms. They are given less time to complete their work. Stress levels are higher.

The more profound sufferers are already finding that they are falling behind their peers at school and they are becoming reluctant to read, as failure has become their natural state. Behaviour may, not surprisingly, deteriorate significantly in some!

The less affected start to have significant difficulties in reading. As these children have had some initial success in their reading, it comes as a surprise, when they start to fail. These children are often criticised heavily, as their difficulty is often perceived as laziness or lack of concentration. If a child brings home a school report stating, "does not concentrate" alarm bells should sound. Teachers must be aware that a high proportion of those accused of lack of concentration may have physical reasons for the problem. There is a duty of care to find out why the child is having difficulties. Obviously some may be lazy, but this should not be assumed.

Migraines and headaches become more common, as stress levels increase. This again will lead to avoidance of situations in which the visual system is extended. Pattern glare becomes more of a problem and, as teenage years approach, hormones affect the neural transmission, reducing the trigger level for difficulties. Depression is often found in those with profound difficulties and anti-social behaviour is common. It is no surprise that such a high proportion of juvenile offenders have dyslexia. The cost of not addressing this problem is high in all ways.

As teenagers develop, they often do not achieve their potential. Those with a limited problem may find that difficulties develop in university, where the workload is such that it becomes impossible to compensate. A stressful job may also prove untenable, without help.

Computers may induce visual dyslexia and migraine, as the stimulus input passes the critical threshold for an individual. Although old research suggests computers have little effect on the visual system, this is incorrect. They can produce a pseudo-dyslexia effect that is extremely unpleasant to those who endure it. Standard optometric testing does not determine this effect and, consequently, is often a waste of time. It can be measured, as can all of the visual effects in dyslexia (using the Optopraxometer)

Panic attacks and agoraphobia often occur in high stress situations e.g. brightly lit shops. Night driving is often a problem, as is motion sickness. Racquet sports are problematical, as following movement of the ball is difficult. In general, adults learn compensatory behaviours to disguise their difficulties.

Chapter 6

Symptoms

Fixation

Fixation problems are common, particularly at a young age, and are characterised by the child having to look away from the page, poor concentration and an inability to maintain focus on an individual word. Their eyes often appear to "bounce" around the page, or they look away, as they read.

Fixation develops with age; at five, many will find it difficult to maintain gaze on a word, at eight, most should be capable – but many are not. Convergence insufficiency, unstable convergence, i.e. the reduction in the ability of the eyes to rotate to a point stimulus, is found and a reduction in the ability to accommodate (focus) is often present. The child will often close one eye, when reading or when tired. The child may disguise this act by laying their head on the desk and covering one eye with their arm, or by abnormal posture or book position.

The treatments for fixation difficulties depend on the cause. The normal optometric treatment (assuming there is no refractive problem) is to use convergence exercises (success is idiosyncratic and regresses when stopped).

Colour prescribing can often stop symptoms, immediately

If the convergence problem is a consequence of suppressional or fusional problems, then it is more appropriate to treat these.

The visual tracking magnifier will often be of great benefit.

Tracking

Tracking is the act of moving the position of gaze along the line of print and from one line to another. If a child has tracking problems, then there is always a visual difficulty. There are many reasons for tracking problems and the treatment will reflect the cause of the problem.

Treatments include colour filters, overlays, occlusion, retinal rivalry modification and exercises.

A further cause of tracking problems is visual plane skew. This is where the appearance of the text is distorted, in such a way, as to make the lines of print appear not to be parallel (e.g. trapezoid.).

Factors that influence tracking difficulties in the classroom include:-

- the font dimensions,
- the distance of the book from the child,
- the lighting in the classroom and
- the physical dimensions of the paragraph.

Proprioceptive problems are common, particularly in dyspraxia.

The child's eyes may have some difficulty, crossing the mid line, and double vision of differing types is common. Changes in text presentation may be beneficial.

Perceptual effects that distort words or letters; or lines changing shape, position or orientation may also produce difficulties, that make tracking difficult.

Inversions and reversals

A common early problem is that of visual reversal or inversions of letters or words. A child often sees words or letters the wrong way round! This common and yet frequently ignored symptom must be differentiated from the sequential memory effect, in which some children see the words correctly, but reverse them within the memory system. Turning the letters or words upside down should be considered visual, in virtually all cases.

Visual reversals

The most common form of visual reversals are d/b p/q and was/saw. The Optimeyes lamp will, by adjustment of the colour of the illumination, often allow the child to see the letters or words flip back and forward, whilst viewing!! Visual inversions are less common, but much more dramatic. The simplest form of inversion is d/q or t/f but in more extreme cases whole words, lines or areas of text can totally invert. Inversions can be treated very successfully by a knowledgeable practitioner.

In both reversals and inversions the symptoms may be found in one eye, only with resultant symptoms depending on the dominance of the eye and the target presented.

Digit span – working memory

If digit span is less than five, phonics will be difficult. Normal span increases by one digit per year, until seven plus or minus two.

Blurring

Blurring of text may be noticed. This may be due to a visual problem e.g. refractive errors, spasm, fatigue or over-stimulus. Alternatively, a visual processing difficulty may be implicated. Professional advice is necessary to determine both the cause and most appropriate treatment.

Blurring is often accompanied by double vision and may, in some cases, be mistaken for double vision, in which the separation of the images is small.

Double vision

Double vision (diplopia) is extremely common in visual processing problems. It may be present in a number of forms.

The most common problem is that of suppression of the double image, created during convergence. This effect may lead to a reduction in the ability of the eyes to converge and can be treated by the use of exercises, colour or prisms. There is much discussion in the optical world on the most appropriate treatment. Although others may disagree, I have become convinced, that colour is usually the best method of treatment, if fusional / suppressional problems are indicated.

Double vision may only be present at one side of eye movement i.e. at one side of the page – colour will alleviate symptoms.

Restriction of eye movement may be present due to this problem and tracking problems may be present. Ball sports may be a problem. It may also be of significant importance in dyspraxia, as proper proprioceptive development may be flawed.

There are more awkward types of double vision. These are found in many with dyslexia.

In some children, reversals of letters may be concurrent with the correct interpretation e.g. a child may see both the d and b

111

combined to create a letter that has a long central line bisecting the circle at the base.

Some words may split horizontally or vertically.

It is often found in one eye only and may be stopped by changing the retinal colour balance.

In some cases, the letter size and type determines the double vision. This may be either a monocular or binocular effect. An annulus of multiple images may be seen or a second image may break and move a significant distance from the fixated image, correlating well with the word or letter displacement, found in so many dyslexic children.

Individual words or letters may also appear to be double. This is a perceptual problem.

Letter or word displacement

Letter displacement may involve the letter sequence being changed within a word e.g. was / saw or the letters appearing to be in an adjacent word or line. Letters may superimpose or disappear. Words may appear in different places, within the text, and, in the most extreme cases, whole lines appear to be in the wrong position.

In rare cases the orientation may be changed e.g. the words may appear at an angle or upside down.

Displacement of words or letters is often seen by dyslexic children and is usually misdiagnosed as a memory problem. If a child sees words or letters in the wrong place, it is always a visual problem and is very treatable. It is absolutely essential that these symptoms are alleviated in any child with reading problems, as soon as possible.

Words or letter vibration

A common symptom in visual dyslexia is letter or word vibration. The most common type is whole word vibration, although single letters may show this anomaly. The vibration speed, direction and amplitude vary.

Word or letter movement

Text becomes unstable and often appears to move spontaneously.

Changes in size or shape of individual letters may be found

In rare cases, the ambient lighting may give rise to characters within text being seen as differing sizes.

Suppression of background

As the eyes converge, it is necessary to ignore the double vision caused by the misalignment of the peripheral retinas. This suppression may be learnt through exercises (with variable results) or it may be possible to suppress immediately by the use of colour. Occasionally, the brain may choose to suppress the central image in one eye, depending on the task. Suppressional problems will result in

- Double vision
- Closing one eye whilst reading
- Tracking problems
- Possible dyspraxia
- Abnormal reading position

Trigeminal nerve effects

The most common symptom is frontal headache or discomfort, that is often misdiagnosed as sinusitis. Migraine can be a more extreme version of this type of headache and is very successfully treated, in a high proportion of cases, using visual techniques.

Hot sandy eyes (dry eyes) respond well to treatment, if inappropriate visual stimulus is the cause of the problem. Hearing can be helped using visual techniques and both the tonal quality and volume can be modified. The ability to filter out background noise is enhanced in significant numbers of people.

Spatial problems

Spatial awareness problems are common and may be due to a number of causes. Areas in the eyes that should be coincident may be displaced or distorted. This inevitably leads to problems with proprioceptive development and may result in symptoms. The most common symptoms that may be found at school are:-

- Dyspraxia
- Inability to catch a ball
- Turning the book or page to an angle, when reading
- Closing one eye, when reading

- Headaches
- Double vision
- Postural changes e.g. head tilt

Visual plane variations

In many children, the visual planes are distorted. This means that, if a child looks at a book, it will appear as though it is tilted, horizontally, vertically or both. The angle of tilt can be significant (40 degrees plus). The floor and other vertical or horizontal objects appear to slope. This slope changes with the angle of gaze.

A high proportion of dyspraxic children will suffer from this problem and, as a consequence, will find their ability to develop spatial awareness difficult. Children may describe the words "falling off the page". The words may also compress at one end, leading to an inability to see the ends of the words, or the letters may overlap. The child may guess the ends of the words.

Mid line crossing difficulties

Children with reading difficulties often experience problems that differ, depending on the position of gaze. The mid line plays an important role in determining the ability of the child to track along the line. Expert advice is necessary.

Fusional difficulties

Fusional problems are often present in children with processing difficulties. These include squint, problems with muscle balance and lazy eye. Treatment of the visual processing difficulty can have significant effects and may alleviate the problem in some cases.

The circle of underachievement

The symptoms associated with visual dyslexia are commonly found, as part of other conditions, and it is likely, that they are related.

It is advised, therefore, that the following conditions are assessed for visual perceptual difficulties, as even if the condition only has a limited visual content, it is advisable to treat, in order to alleviate the symptoms found:-

- Dystonia - in particular blepharospasm
- Dyskinesis – in Parkinson's disease
- Depression – it appears that seratonin levels may be changed for some
- SAD – seasonal affective disorder may benefit even more
- ADHD – a high proportion of those with attention and hyperactivity difficulties respond quickly to light treatment
- ADD – attention and concentration may be influenced by light input.
- Autistic spectrum disorder should be assessed, if possible, as a high proportion will benefit.
- Aspergers syndrome – most will have visual perceptual difficulties
- Migraine – about two thirds of sufferers will benefit from an assessment
- Trigeminal neuralgia (including misdiagnosed sinusitis)– will often respond to visual treatment
- Menieres disease and tinnitus sufferers can benefit
- Field defects – will occasionally reduce with treatment

- Strabismus – can sometimes be rectified by using colour. If successful, this is the simplest and least disturbing treatment

- Anisometropia and anisokonia – unequal strength lenses or eyes with unequal image sizes (more common than most opticians think) will often be significantly improved by colour treatment

- Amblyopia – lazy eye treatment is often very successful (and immediate), when using colour

- Anti social behaviour and truancy – is the child saying that the visual stimulus in schools is causing them problems? Frequent truants and those exhibiting antisocial behaviour are often sufferers of visual stress.

Chapter 7

Assessment

There is no doubt that a significant proportion of people do not achieve their maximum visual processing potential.

A standard eye examination will not usually assess visual processing difficulties. Consequently, it is essential that the limitations and benefits of an eye examination and where this may be used, to beneficial effect in visual processing difficulties, are understood. A visual processing difficulty will not usually be an organic visual problem, as it is not related to the strength of the lens that is prescribed in spectacles, but, as it may give similar symptoms, it is necessary to discuss the problems that an optometrist will treat. So what is the current role of the optometrist?

The eye examination

To understand the need for an optometric examination and the limitations imposed by an eye examination, it is essential that basic knowledge of optometric procedures be given. The legal status of the optometrist varies from country to country, but, generally, his role would be to detect any eye or systemic disease, that manifests itself within the eyes, detect and prescribe for refractive errors, detect and treat muscle problems, fit contact lenses and dispense spectacles. It would be unusual for him to detect and treat visual processing deficiencies, as part of normal responsibilities. So we need to know a little of the optometric routine and its relevance to visual processing difficulties.

In an eye examination, the optometrist will always have asked, whether there are any visual problems. He will be particularly interested in symptoms of eye disease, such as glaucoma, but will place less importance on those of processing problems. It is not unusual to discount symptoms, that are not related to disease or muscle anomalies. In visual processing, symptoms that are produced during reading text are of particular importance. These include:

- Headaches
- Rubbing of eyes
- General visual discomfort
- Double vision

- Seeing colours in text
- The print appearing to flash on and off
- Words moving around the page
- Jumbling or vibrating words
- Inability to track along a line, correctly.

There can be many reasons for the above symptoms and the optometrist will try to determine the aetiology (cause) to enable resolution. It is unlikely that visual processing will play a great part in his deliberations.

He will then measure the refractive correction (i.e. how strong a lens is required) and the ability to discriminate detail (i.e. the ability to see small print). Accommodation (focusing) and convergence of the eyes will be assessed. Sometimes, he will assess the ability of the eyes to integrate and fuse their slightly different perceived images. The health of the eyes will be assessed and eye movement will be examined. The optometrist will often make a judgement on the relevance of any discrepancy from the norm and take action, as appropriate. It is unlikely that signs of visual processing difficulties found in the eye examination will be considered of clinical importance. Signs found by the optometrist in an eye examination, that can indicate visual dyslexia, include

- Accommodation reduction,
- Convergence insufficiency,
- Visual saccade and visual pursuit problems,
- Photophobia, visual acuity reduction, fusional reserve deficit
- Suppression anomalies.

We will have to examine all of these in detail, if we are to understand visual dyslexia.

Accommodation reduction

The process of accommodation is the ability of the eyes to change the shape of the lens, within the eye, to allow clear vision of a near object. In other words, the lens changes focus, as we bring text closer. It can be compared to the focusing of a camera lens. Accommodation reserves(i.e. the amount of focusing range, available to be used by an individual) reduce throughout life and are a predicator of age. For instance, if you have perfect distance vision, you will need reading spectacles when you are in your mid 40's, (earlier for women than men, because the former hold books closer). In a child with reading difficulties, this ability to focus will often be markedly lower than expected. It may exhibit the focusing ability of someone 20 to 30 years older. This reduction in focusing ability is immediately treatable, using filters, and, in most cases, major levels of improvement are attainable. Standard optometric treatments are often unsuccessful. The conventional treatment is to use a reading addition i.e. a spectacle correction that compensates for the loss of focusing. This does not increase ability, but a properly prescribed filter will and, it follows, that an incorrectly prescribed filter may be detrimental.

Convergence insufficiency

Convergence is the rotation performed by the eyes, when looking at a word. The degree of rotation is dependent on how close a word is to the eye. It also varies as fixation changes from one side of the page to the other.

The eyes have to rotate inwards, to fix on a near point. It is a learnt process and many children are slow to develop this skill.

In a child with reading difficulties, this ability to converge may be reduced. It is often particularly noticeable, when focusing quickly, from a far object to a word in a closely held reading book (this is called jump convergence). In a classroom, this is the same as looking from the board in the distance to a book on the desk. An inability to perform this action satisfactorily is treatable, with significant improvements possible. The standard optometric treatments are exercises, or prisms incorporated into spectacles.

Exercise results are variable and require great perseverance; prisms will allow the eye to rotate less, when looking at a close fixation point. Less commonly used are the processing techniques using filters, but their results are often better. This raises the question - if filters will reduce the convergence problem, do the lights in the classroom create or intensify the difficulties that many children have in dyslexia. The answer is yes; therefore, lighting must be a consideration in treatment of visual dyslexia.

Saccadal Problems

A visual saccade is the movement made by the eyes in focusing In the visual saccade, the child has to fix on one word and, then, move along the line and fixate on a subsequent word. The visual saccade also involves small corrective movements. In a child with visual dyslexia, there are more visual saccades, in a reverse direction, than normal. Children with visual dyslexia will often have difficulties tracking along the line of text. They will often jump a line or they may find that they have lost their position in the text. As they move from one line to the next, they have difficulty moving the correct distance down the page. This is a binocular vision problem, in most cases and is usually treatable, relatively easily, by using processing techniques. Standard optometric tests ignore this problem and do not treat it. It is quite likely that a child will also have problems with pursuit vision. Although not relevant in reading, it does have a marked effect on sporting activities, particularly racquet sports. The child will usually have difficulty in determining the position of the ball.

Photophobia

Photophobia may be defined as the dislike of light. Many people with visual dyslexia have significant and often debilitating problems, due to bright or flickering lights. Some find that it proves to be extremely difficult to go out of doors without sunspecs; some, have permanently screwed up eyes and go around the house turning the lights off. Photophobia, in its most extreme form, involves reflex tearing and significant discomfort. Tinted lenses would be prescribed, in conventional optometric practice, although, the colour and the density of the tint will be a matter of guesswork, in most cases. In a visual processing assessment, this can be prescribed with confidence.

Visual acuity reduction

We often find that one or both eyes does not see, as well as expected, even with the "correct" refractive spectacle prescription. This can be due to visual processing problems. Modification of stimuli, by reducing an overloaded system or attempting to force the system to work, can have beneficial results. Visual acuities often respond, immediately, if stimulus reduction is necessary, thereby making it difficult to assess how we define acuity. Optometric testing of visual standards

127

has to be questioned, as the method currently used is unreliable. Visual acuity is a poor indicator of the maximum visual standard attainable.

Fusional reserve deficit

In children or adults with visual dyslexia, the ability, for the eyes to bring their images together, is often impaired. There are also monocular fusional anomalies, that appear in monocular pattern glare and the reversals, shown in the Jordan Reversal and Integration test. These fusional problems are often associated with other conditions, e.g. migraine, strabismus and amblyopia. The central area of binocular fusion is often smaller, in those with visual dyslexia than those with "normal" vision. Filters will often have an immediate effect.

Suppression anomalies

Often, the child with visual dyslexia may experience difficulty in suppressing unwanted information, being sent from the eyes to the brain. Physiological diplopia (that is double vision due to anatomical features) is often present. This type of double vision is normal and is usually ignored.

Squint

Squint is often present. This involves a suppression of vision from the central area of one eye, which then converges or diverges, to allow binocular stabilisation of the peripheral vision and comfort of the non-squinting eye. Alternating suppression, from one eye to the other or from the nasal to the peripheral side of either eye, can be very disturbing. Some suppression problems can be treated simply; others are much more difficult and treatment can have unpleasant side effects. It is unlikely that processing techniques will be used, at present, to treat suppression problems, although it is reasonable to expect that the professional should know, when they are the most appropriate. Not many optometrists will be aware of these techniques. They are often the least disturbing methods, of all, for modifying suppression.

A multi-faceted input

Many problems with dyslexia may be considered to be a response to sensory overstimulation. As well as the visual and auditory innervation, olfactory, tactile and vestibular inputs may also have an effect. The point, at which difficulties occur, is also determined by a number of other factors, such as diet, the hormone levels of the individual, allergic responses, oxygen and lactic acid levels and any medications or drugs being taken. The relative importance of all the input factors may vary, from one person to another, and some may be more sensitive to different stimuli or combination of stimuli. For some, dyslexia may be considered as an adverse response to the input stimulation and other factors may amplify or reduce the effects of the combined stimuli. The level of stimulus input, that results in problems, is described as the trigger level. This is variable, depending on input type and body chemistry.

As the input stimulus increases, there is a point, at which most people will exhibit symptoms. In those with visual dyslexia, or visually triggered migraine etc., this point is approached, too quickly. Therefore, to reduce these difficulties, it is prudent, either to reduce the stimulation levels or, alternatively, to change the body chemistry by removal of any chemical that effectively makes the trigger easier to hit. Some foodstuffs contain chemicals that make these triggers easier to "hit" and elimination will reduce problems.

The symptoms diagnose the condition. Now, however, it is possible to quantify visual dyslexia experimentally.

The visual processing expert

The obvious person, that should undertake a visual processing assessment, is either the optometrist or the optician.
For the optical professional, the prescribing of spectacles and contact lenses must now be considered as much more than just putting lenses into the frame to correct any refractive error. We have to consider the effects of magnification; we have to consider the effects of visual destabilisation. What are the effects on visual processing of the spectacle prescription, the way it is made up and whether there is a more appropriate treatment? Opticians will have to think about the effects of lenses; they should think about the effects of stimulation of fusion. It is difficult to see how somebody, without a significant level of knowledge, can do this job properly.

The dangers of poor dispensing in visual dyslexia

Sadly, in a lot of places, people with limited knowledge dispense spectacles. I personally find this of great concern, as there is a massive difference between the best possible and the basic (and, sometimes, inadequately considered) prescribing. However the latter is often legal, as long as an optician is in the practice. One must be fortunate to find an optician, who has thought through all the potential problems, as his training in visual processing is often limited, at present.

It is very difficult for an optician to know the full visual processing background of a person, without extensive questioning and assessment. Self-selection is not sensible and "shopping around" may be foolhardy. Doing that, risks your child's abilities and, if there is any suspicion of problems (e.g. either squint or areas associated with dyslexia), it is essential to spend a lot of time with the optician or doctor to explore all the particular problems that could be associated with it. A 'quickie' test is not satisfactory.

A visual processing assessment using the Orthoscopics system.

The Orthoscopics system is what I consider to be the state-of-the-art assessment for visual dyslexia and other perceptual problems. There are other techniques, but the Orthoscopics system represents a quantum step forward.

After the eye examination (and orthoptic assessment if necessary) the Orthoscopics assessment can prove to be a life changer.

A comprehensive history and symptoms is taken, followed by observation of gait, posture and speech. The Read-Eye instrument will allow the most accurate colour assessment possible, the T-test will allow for plane skew, the Optopraxometer will assess the effects of lenses (including the HOYA V-DEX range) and the Optimeyes will allow assessment of postural and vestibular effects. Immediate and significant improvements will, usually, be found and, normally, a 6-10 page report for the school would be provided.

For those that do not have a visual component or exhibit other symptoms, in their difficulties, the patient would be referred to the appropriate person for further investigation

For more details www.orthoscopics.com

Chapter 8

Treatment

Visual dyslexia can be considered to be the result of a number of different conditions and, therefore, can be expected to respond to a number of treatments. However, the differentiation of these conditions may be difficult and it is likely, that there is more than one problem area for most sufferers. Many anomalies will respond positively to more than one regime and, therefore, it is impossible to say, which is the best treatment for any particular type of visual dyslexia, without comparative research being undertaken and the results being analysed. This comparative research has not been undertaken, to date. The professional who states "this is the only treatment that can work" is, at best, naïve and is expressing an opinion, which may be unhelpful to the possibe treatment methods of which he has no knowledge. So, how do we access the most appropriate method of treatment?

Due to the difficulties in assessment and the relative lack of comparative treatment research, it is perhaps most realistic to consider a scattergun approach and to try to maximise the benefit, by utilising more than one treatment. We will therefore go through the principal visual treatments, as well as some other, less common, physical treatments, all of which seem to help the visual processing system.

As visual dyslexia consists of a number of disparate conditions or effects, two questions must be posed:
"Can all types of visual dyslexia be categorised?" and "Does the category affect the treatment employed?"

To enable treatments to be assessed, we must have a form of differentiating the different types of visual dyslexia. Inevitably, this differentiation is arbitrary and, to some extent, subjective. I have therefore made a suggested list of types of visual processing deficit, that may be grouped together as the types of visual dyslexia.

- Monocular fusional dyslexia
- Binocular fusional dyslexia
- Suppressional anomalies
- Retinal wiring anomalies

- Over stimulation effects
- Visual sequential memory
- Inhibitory distortions
- Outside stimulus effects
- Vestibular effects
- Trigeminal effects
- Magnocellular and parvocellullar processing difficulties

Monocular or binocular assessment and treatment?

It is now clear, that there is a monocular element as well as a binocular component in visual dyslexia.

This can be shown in pattern glare tests, the Jordan reversal test and, sometimes, in colour and brightness variability between the two eyes. It is therefore essential, that a monocular assessment be performed, as well as a binocular assessment, so that a decision can be made, as to whether to treat each eye individually, or whether to consider a binocular treatment, as being the best way forward.

Assuming that a monocular anomaly is found, then there are a number of monocular treatments that may be used. The most common is that of occlusion, i.e. covering the "offending eye", with a patch, or using a semi-occlusive method, with frosted glass being placed in front of the eye which is creating the problem.

This treatment may be utilised when the Dunlop test (a test for the separation of the position of gaze of the eyes) is used and fixation breakdown is observed. The clinician attempts to improve the ability of the child by creating a leading eye, in order to develop fixation and, as a result, visual stability improves.

A simple tracking test will also suggest monocular assessment. If a person covers one eye in reading a passage of text, it is often apparent that improvements occur in the traversing of a line of text. These movements along the line are often called "tracking movements" and consist of forward and reverse saccades. Therefore it is a worth experimenting with the covering of each eye, in turn, to determine whether occlusion may help. If this technique is helpful, it is essential that the subject seek professional advice, to allow a full evaluation of the binocular function in both static and dynamic situations. It may be that occlusion is the preferred course of action, but it must be stressed that this treatment may have significant potential problems in binocular development and, therefore, whilst it is safe to try for a few minutes, it must not be attempted as a long-term treatment, without professional involvement.

A variation of this treatment is the using of filters or tinted lenses in spectacles or incorporated in contact lenses. This technique, although relatively rarely used, can be very successful.

Monocular filters

Filters are assessed, individually. They are prescribed in either a contact lens or spectacle form that may be cosmetically enhanced, on spectacle lenses, using mirror coatings. The contact lens may be fully tinted or may contain a clear or differentially tinted area. The tinted contact lenses, that are suitable for visual dyslexia, have also been used in colour vision problems, for a considerable time, with varying degrees of success. In general, I have found better results with red/green colour deficiencies, than with other colour deficiencies, whereas the blue lenses appear to be more successful with visual dyslexia. However, spectacle lenses can be prescribed much more accurately and would normally be preferable

Virtually all tinted spectacle lenses, presently used, are broad-spectrum filters. This means that they have a fairly uniform transmission of light through the visual spectrum and do not completely attenuate any wavelengths of light. Do not expect the same colour of tint to be prescribed in spectacles as contact lenses as their physical properties are different (this also applies when comparing the effects of overlays and tinted spectacles). I will review the effects of filters, in greater detail, later.

In rare cases, I have found it necessary to produce different effects on the peripheral retina, relative to the central areas. In other words, only part of the retina has a filter placed over it by virtue of the contact lens remaining in the same position relative to the back of the eye and moving with eye rotation. This technique is only possible using contact lenses and cannot be undertaken with spectacles. Tinted contact lenses, designed to combat visual dyslexia, usually make the pupil size appear larger in the eye that contains the lens, with occasional unusual cosmetic effects in bright lighting conditions. There are particular fitting criteria for these contact lenses in order to minimise movement and reduce the effects of displacement. A high degree of expertise is required to maximize the effect of tinted contact lenses.

There is an interesting side effect of monocular tinting. This is the Pulfrich phenomenon. This curious effect occurs, when one eye has a neutral filter placed in front. A pendulum that is made to swing from side to side appears to rotate in an elliptical path rather than swing in a straight line. It has been suggested that this may cause problems with spatial orientation. However, in every case of a patient whom I have fitted with a monocular filter, this effect does not appear to have been noticed and, I suspect, that it is insignificant, in practice.

I believe that I pioneered the use of contact lenses in practice, for visual dyslexia and their use for partial occlusion. They are of particular use, when a child will not use spectacles, for vanity reasons, and where a fusional binocular problem is suspected. There are drawbacks which cause me to fit tinted lenses in spectacles in preference, although I always maintain the option of fitting contact lenses.

Contact lenses for image size differences between eyes

Contact lenses are the first choice, however, where there is a significant prescription difference, from one eye to the other, and the person's areas of fusion do not correspond at a retinal level, due to the relative prismatic (the movement of position of the text, due to the thickness of the lenses in the spectacles) and magnification effects of the spectacle lenses. Prismatic effects are negated, due to the minimal differences in contact lens thickness and, effectively, the vergences i.e. the changes in direction of the light, due to the thickness and shape of the lenses, are parallel.

When the eye rotates, the relative prismatic effects become too great to fuse, often causing difficulties, which give similar symptoms to those experienced in visual dyslexia. Therefore, prismatic effects have to be addressed in spectacle dispensing, when there are reading difficulties. This is an extremely complex area and requires specialist knowledge.

Magnification differences, between the images formed on the retinas of both eyes, can be a problem leading to fusional difficulties, in cases of severe astigmatism, and, in rare cases, to retinal fusional disparity.

Where there are relative magnification problems, it can be shown that a prescription difference of about four dioptres between the eyes would normally be at the limit, at which spectacles would allow fusion. However this is a generalization and four dioptres may be too great a retinal magnification disparity to be acceptable to some. Contact lenses must be the preferred choice in magnification differences.

The provider of spectacles should be aware of these potential problems. Sadly, most do not understand, ignore or prefer the risks of suppressional responses, to that of explaining that there are alternative treatments. It must be accepted that all treatments have disadvantages and that clinical considerations are necessary in contact lens fitting

Small optometric prescriptions

It may be, that extremely small changes in power may be of great significance for some people. It may not be a coincidence, that the most common prescription, in children having visual dyslexia symptoms, is often a very small plus prescription, whereas, it is rare to see a very small minus prescription. A reading problem, with a small hyperopic correction, should always be considered suspicious, particularly if there is a reduction in visual acuity in either eye. This power sensitivity may be due to the amount of accommodation variability, as the eyes cross the mid line, although there may be other explanations. Therefore, it may be sensible to try spectacles with very small prescriptions, if there is a tracking problem. It should be borne in mind that, in the end, the parent is the most appropriate person to make an informed decision and that the professional should endeavour to make the parent aware of all the pros and cons of any particular treatment regime. The professional is not always right!

Differential filters

When spectacles are tinted, it is sometimes better to tint one lens only, or to tint each eye, individually. The cosmetic reaction is often better, with two lenses tinted equally, and judgement is required as to the best way forward. It must be remembered that if the cosmetic option is to use a mirrored lens to equalize the cosmetic effect, then the combined tint has to be assessed, i.e. the mirrored surface has an effect on the filter prescribed. In many cases, this is not an appropriate method and other options may be better.

When using a monocular tint, we also may have effects on retinal rivalry, although constancy of vision appears normal. Research is scant at present, but, from the hundreds of patients I have seen who have undergone this treatment, it appears that there are no detrimental effects.

The success rate of occlusive methods is high and improvements are immediate.

Cutting out peripheral vision

Two other variations in occlusive techniques are that of using a pinhole or stenopaic slit.

A pinhole, placed in the centre of a dark spectacle lens, immediately stops the peripheral distortion and the amount of light is reduced significantlyl. A greater depth of focus is present, due to the pinhole camera principal. Pattern glare usually stops, immediately.

However, this treatment has many disadvantages and is usually best when used in diagnosis, although it may have a part to play in screening.

A stenopaic slit is an elongated pinhole, that may be rotated through ninety degrees. Pattern glare appears to be angle dependent, i.e. as the slit is rotated the pattern glare effects are modified or stopped, and, as the orientation of the text is changed, pattern glare will also change or be modified.

This effect can be demonstrated by rotation of the slit to different orientations. The 90 and 0 degree positions are most common in being the best and worst for different individuals, although either position can be best.

Recently we have introduced a variation of the monocular occlusion, by using polarized neutral filters. A neutral filter is a sheet of plastic that does not affect colour vision.

If a person looks at text and the polarized filter is rotated, coloured visual effects, that appear to flash, may be produced, or there may be a change in the perceived size of the text. On others, the rotation appears to have an effect on visual acuity. The mechanisms that give rise to these effects are unclear, but it may be surmised that polarized light may have an effect on visual processing. This may be one of the reasons, why some print creates difficulties, in cases of reflection from a school whiteboard or from some types of paper.

In summary, occlusion is a particularly good treatment in certain situations, but partial occlusion seems to hold promise of more improvement. The question is, which is the most appropriate type to use? In general, it is perhaps better to rely on the responses of the person being treated and to allow them to make the decision.

Advantages of monocular occlusion
- High success rate
- Relatively easy to screen

Disadvantages of monocular occlusion
- High degree of optometric knowledge required
- Potential binocular anomalies may be introduced
- Contact lens knowledge is required

Filters and tints

Although filters and tinted spectacles have been used since the 1930`s, for the treatment of visual perception anomalies, which included visual dyslexia, they have not been used by many practitioners. It is just in the past twenty years, that it has become acceptable to use filters or tinted spectacles, in the treatment of visual dyslexia.

All tints or filters reduce the amount of light reaching the eyes, either by absorption, reflection or destructive interference. The most commonly available filters are broad-spectrum absorption filters. These are principally produced in plastic, for ease of absorption tinting, although the range of tints, whilst appearing extremely large, is restricted by the matrix of the material. These filters may appear to be specific in the colours of light they allow to be transmitted, but their transmission curves are such, that the wavelengths attenuated (stopped) are fairly diverse, in most cases. Commercial claims may be dubious, as to the precise need for a specific filter range, as the light transmitted varies, depending on the environment and the light source.

I find it untenable to state that only one precise (although broadband) filter is acceptable for a given child. In practice a fairly wide spectrum of filters may be necessary depending on light source and task. In practice, I have found this to be the case, bearing in mind that different lighting and task demands may require changes for some, although pragmatic compromise is acceptable.

It would also be wrong to assume, that a single filter is perfect in all situations. This is impossible, as different lighting conditions may have massively different characteristics, and it is unreasonable to expect a single filter to be totally suitable. It must also be understood that different lights can have different effects, dependent on their position in a room, their flicker frequency (the amount of times per second that the fluorescent tube flashes), the modulation of flicker (the difference in brightness between the on and off state) the state of maintenance of the light bulb and the operating temperature. Daylight varies by a remarkable extent, ranging from dusk conditions to very bright sunlight.
It is impossible to guess the transmission of a tint, by its colour, as this can vary wildly, as a function of the material or dyes used. To state that a person needs a blue lens is misleading, as the characteristics of different blue lenses can vary wildly.

There was a particular example of the futility of attempting to achieve the same effect on a patient, by matching filters visually, which underlines the need for a better understanding. Approximately fifteen years ago, a manufacturer discontinued the production of a blue glass lens filter. This filter was used frequently and, when people were replacing their spectacles, an equivalent plastic lens was used, with a colour that appeared to be identical. The transmission curves were however significantly different although, at the time, relatively little thought was given to this. The prevailing attitude was that, apart from colour changes induced by the transmission curves, there were no significant problems in using any filter. Coloured lenses were not supposed to have any effect. This complacent attitude was abruptly changed, by the resulting level of problems, which people had with these "equivalent" lenses. It is only recently, that it has become apparent, that the assessment and accurate prescribing of tints have a direct effect on everyone wearing them, not just those with dyslexia. Optical professionals will have to prescribe filters from a position of knowledge, rather than just hoping for the best. The status quo is unacceptable!

There are a number of problems with the normal types of lenses, used in the treatment of dyslexia. Usually these are plastic lenses, (generally CR39) which have the disadvantage in that the absorption of the dye colour is determined by the pore size within the material. The matrix used does not allow large molecular dyes into the material and, therefore, it is difficult remove the red end of the visual spectrum. This is particularly important, in that red is often the most disturbing colour to many of the people with the problems of dyslexia, migraine and epilepsy. All absorption tinted CR39 lenses allow a disproportionate amount of the red end of the visual spectrum to be transmitted, regardless of their perceived colour. It is pleasing to note that new filters (HOYA V-DEX) have reached the market and can stop significant amounts of red light passing through the lens.

It is possible to filter out blue, green and yellow, more successfully, using CR39 lenses, but it is rare to attempt full absorption, as the lenses tend to be too dark.

Other filters in the plastic range tend to be broad-spectrum filters, in that they don't filter out specific wavelengths of the visual spectrum, very selectively, and they just generally reduce the transmission in certain parts of the spectrum. It must not be forgotten that any filter reduces light transmission and that there may be situations, in which the tint may not only be unnecessary but potentially dangerous. Caution is advised as to when to wear specific filters, as it is perhaps not advisable to use a dark tinted lens, under poor lighting conditions or at night. Discretion must be used, when driving at night, and legal requirements must be met. Colours of everyday objects may be found to be modified by highly specific filters and this may require action, in some situations. However, the person can usually cope with these effects, as there is a mechanism that maintains colour constancy in the visual system. Anti-reflection coatings may be utilised on the lens, as a method of increasing clarity, and some people prefer lenses with filters that reduce ultraviolet transmission, although it is unclear as to whether these have a relationship with the dyslexic or glare producing areas.

Tinted lenses have a number of effects that are both positive and negative.

Positive effects of tinted spectacle or contact lenses of correctly prescribed filters.

- Increase in accommodative reserves, i.e. the eyes can focus on a nearer point, than without the tint

- Improvement in convergence ability, i.e. the eyes will rotate to a nearer position. This is particularly true, when jumping from a distant object to a near object, such as in looking from a blackboard to a word on a page.

- Improvement in visual acuity, i.e. an improvement will be noticed in the ability to see small print. It must be pointed out that the best filter may not be the same for distance and reading.

- Increase in fusional reserves, i.e. the ability to fuse the different images produced by each eye will be enhanced. The area of clear central vision appears larger.

- Improvements in reading accuracy and speed will be noticed. In some cases, the reading speed is reduced, as scan reading techniques are modified. This allows the full word to be seen, not just the first letters and, consequently, the scanning is slower but more accurate.

- Pursuit and saccadal eye movements are more controlled. This allows better tracking of moving objects and the refocusing on reading text is improved

153

- Pattern glare will be eliminated. The effects of environmental patterns or reading text will stop producing unstable vision and the unpleasant visual hallucinations will be eliminated. This is of particular importance in the stopping of frontal headaches and migraine.

- Visual discomfort is reduced / eliminated. Hot uncomfortable eyes, characterized by eye rubbing, will be "calmed".

- Migraines or headaches will be reduced or eliminated. All headaches and migraines are not visually provoked, but the high percentage that are, will be helped immediately.

- Hearing may be improved. Although the mechanism is unclear (the trigeminal nerve appears to be implicated), tone and volume seem to be affected by the convergence reflex and, as the reflex is modified by the improvement in convergence possible with the use of a correctly prescribed filter, the resultant improvement in hearing is immediate.

- Balance may be improved. The vestibular effects of a correctly prescribed filter are such that there is an immediate improvement in balance, in a high proportion of cases.

- Sequential memory may improve. A number of people have reported improved sequential visual memory. Although these reports are anecdotal, there is no reason to doubt their veracity, but not every person with tinted lenses reports this effect.
- Fixation may be improved.

Disadvantages of tinted lenses

- Increase in pupil size may allow unwanted ultra violet into the eyes, if a suitable inhibitor is not used. However most plastic lenses do reduce the transmission of ultraviolet light.
- Iris sphincter muscles may not contract, as much as is desirable, thereby losing tonus of the muscles that alter the pupil size in the iris.
- Wearing a tint that is too dark may cause problems in poor lighting conditions. In night conditions, it may be that a person reduces the light entering the eye to such an extent that their visual performance is diminished markedly.
- Specific filters will alter colour rendition. In some particular cases the colour changes may cause difficulties e.g. in dye matching

- There may be legal restrictions on the use of tints or filters, in some circumstances
- Cosmetically, some will object to tinted spectacles. This applies for example in cases where the colour is the "wrong" one for a boy who supports a particular football team and the prescribed filter is the colour of a rival team. Great anguish can be caused!

Visual sequencing of letters and phonics

In many with visual dyslexia the visual memory is poor and recall of letters or sounds is difficult. This confusion in the order of things, creates many problems in reading. It is important to realize that digit sequential memory develops with age, about one digit per year, up to seven years of age and then stabilizes. A range of recall of five to nine digits is considered normal, in a subject over seven years of age, although wide variations can be present.

A reduction in recall memory to two or three digits is common and can have debilitating results. If sequencing problems are present, it may be impossible to use phonics until the sequential memory is satisfactory. If recall is less than four digits, it will prove very difficult to use phonic teaching methods. It is therefore necessary to increase the memory ability of the child, to enable him to remember the start of the word, when he reaches the end.

It is possible to effect sequential memory improvement changes, by using filters or memory switching or repetition. Allergies and sensitivities must be addressed and anti-histamines may well help. Recent research has shown, that fatty acid supplements are helpful, in some with problems. These fatty acids are essential in neuro-transmission and, in a proportion of those with symptoms, they may be of significant help. They may be of greater assistance in those with a tendency to allergies or sensitivities, although the research is still incomplete.

Drugs

Drugs, designed to alleviate the symptoms of migraine, will often help dyslexia as well as migraine sufferers. Other drugs may also be beneficial or cause greater problems. Motion or travel sickness remedies are often helpful.

157

Little research has been undertaken on drug treatment, to date.

In some cases drugs may be variable in their efficacy e.g.

Ritalin will often be beneficial, at first, but, after a sustained

period of treatment, often has detrimental effects.

Homeopathic medicine seems to be helpful, in some cases.

Lighting

Lighting can provoke the symptoms of visual dyslexia.

There are a number of types of stimulation involved in lighting:

- brightness,
- hue,
- saturation,
- flicker frequency,
- flicker modulation.

These must all be combined with the other stimulus inputs,

such as:

- pattern,
- sound,
- vestibular,
- olfactory
- tactile,

to create an overall stimulus. Sensitivity to visual stimulation is determined, not only by the visual stimulus, but also by all other stimulus inputs and the physical state of the body.

It can be shown that modification of any of the stimuli may have positive or negative effects.

.

Rote learning and visualization techniques may be of assistance. Alternatively, computer programs such as 'Brain Builder' can also be used. Significant levels of improvement can be made in the memory of a person, by using these techniques and this can make a major difference in phonic sequential processing.

Diet

In some with visual dyslexia, there is almost certainly a dietary aspect to the provocation of visual dyslexia.

Some foods are more likely to cause problems and it may be a wise move to consider elimination or reduction of certain foods. The child may suffer from headaches, stomach aches or hyperactivity. Attention deficiency may also be noticed. The most likely foods to cause problems are common and, consequently, it may be prudent to remove each from the diet, experimentally, to determine the likelihood of sensitivity.

Chapter 9

The Cambridge trials

September – December 2002

In September 2002, a team was brought together to look at the physical aspects of visual dyslexia, to see whether there were measurable physical effects and to prove or disprove measurability of the condition.

The results were dramatic, conclusive and have far-reaching consequences. They conclusively showed that it is impossible to rely on any educational assessment, unless visual stimulus is tested FIRST.

The study was financed by HOYA lens UK Ltd, with help and co-operation from Norvilles Ltd, Diverse technologies Ltd and BPI (all well-known optical companies)

The trial consisted of two parts

    1 Optometric

2 Psychometric

The optometric part of the trial consisted of an eye examination, a questionnaire, a colour space analysis and measurements of performance both before and after prescribing, using the optopraxometer with and without prescribed lenses.

The psychometric part of the trial used standard psychometric tests, plus EEGs and quantitive balance tests.
The children to be tested ranged from 8 – 12 years
There was a control group. The children with reading difficulties were self selected on a "first come" basis, in response to an advertisement.
In all there were 50 initially selected for the reading disabled group but a small number were ruled out for various reasons e.g. pathology

Optometric results

Apart from 4 individuals, all children would not normally have been prescribed glasses as a result of a standard eye examination. Virtually all were slightly hyperopic, which is normal. A high proportion showed exophoria (difficulty with converging) at reading distance and some showed variable muscle balance difficulties.

Symptoms commonly included frontal headaches, sore eyes, differing types of double vision, displacement, reversal and inversion, tracking and fixation problems, concentration difficulties, flicker sensitivity, pattern glare etc.

About 80% of children had a significant number of visual perceptual symptoms.

Colour space analysis was variable but the resultant lenses were interesting, 80% of the lenses prescribed for classroom conditions were unable to be prescribed, prior to the VDEX range being developed. The Optopraxometer and Read-Eye combination was effective in determining veracity of results and our confidence level was very high.

The Optopraxometer results were impressive.

- In the three convergence tests, the AVERAGE improvement was 6 cm nearer the eye,

- In the tracking test, the average improvement was over 4 cm

- In the accommodation test, the average improvement was 5cm

- In the central vision stability test, the average improvement was 7cm

- In full stability, the improvement was an average of 10cm!!

These results are dramatic and show that inappropriate lighting can have detrimental effects on the visual system. At last, a quantification of visual dyslexia was possible.

.

Two things also came out of the tests. Symptoms ceased with the correct lenses and, the worse the Optopraxometer initial position, the greater was the improvement. Many children could not possibly have read, with their level of visual perceptual problem, and it became clear that it is essential to correct this anomaly. Educational interventions must be preceded by perceptual correction.

Concentration could be considered an indicator. The greater the concentration problem, the greater was the visual perceptual difficulty (not surprisingly!). A school report that states that "a pupil does not concentrate" should set alarm bells ringing.

Psychometrics

A battery of psychometric tests was performed, under school lighting conditions, and it was clear that there is a visual influence on the results.
Hearing, phonics and understanding were correlated with convergence difficulties and a strong link was established. With appropriate visual treatment, there was a significant improvement in what has up to now been considered a language problem. A visual link had now been established.

Balance

Quantitive methods were used to compare the posture and balance of the children. It was found that the dyslexic children had significantly poorer balance, than the control group, under "school lighting" conditions, and that this improved, significantly, with correctly prescribed lenses. However, with randomly prescribed lenses, the control group balance became

165

worse, indicating the possibility of provoking symptoms with inappropriate lighting or lenses.

EEGs

An EEG measures the electrical patterns inside the brain by means of connections attached to the head.
Nineteen point EEGs were performed on the children.
The results can only be described as "mind blowing"
The dyslexic children had a markedly differing EEG under school lighting, when compared to the control group. It showed the occipital cortex appearing similar to the control group WITH EYES CLOSED. In other words the brain was in a "switched off" mode, with increased alpha activity in the occipital and surrounding areas. This pattern was normalised by using correctly prescribed lenses. The control group did not exhibit this neurological effect.
The implications are important for a number of reasons:

- All current tests (educational and psychological) for dyslexia are potentially flawed, if the lighting is not optimal for the child.

- The child should expect that the visual environment is acceptable to his processing needs
- All optometric tests, in a non controlled lighting environment may be inaccurate
- Examinations results may be reduced for a student thereby changing his or her life!
- Concentration problems may be provoked
- Dyslexia may be provoked

A new assessment order is necessary.

In future it will be unacceptable for a child to under-perform at school, without a visual processing assessment being undertaken. All other tests performed should be subsequent and in addition to the visual processing test.

Chapter ten

A new model for Visual dyslexia – the dynamic model

Let me introduce a new model, that can explain the symptoms in visual dyslexia and other conditions

Introduction to previous models

The symptoms that have been discussed, previously, have been noted by many in the past. Many have been grouped together and described as the Irlen Syndrome or Meares Irlen Syndrome. There have been a number of theories as to the aetiology of the symptoms, and treatments have been prescribed, to fit in with theories that have been proposed. These treatments have had varying degrees of success and, although their exponents often claim their methodology is the only one that can work, it is clear that this is not the case and that more than one treatment will often be successful, for a given person.

The old theories, as to why people have visual perceptual difficulties, fall broadly into four camps:-

- magnocellular dysfunction
- overstimulation
- developmental delay
- attentional problems

However, none of these theories gives a good match to the symptoms described by the person, with difficulties, and, consequently, I will speculate on a model that appears to match virtually all symptoms described – the Dynamic model.

Before describing my model, I will briefly describe each of the previous models.

Magnocellular dysfunction

Dysfunction of the magnocellular system has been suggested as the model for visual dyslexia. It suggests that peripheral vision dysfunction is caused by small magno-cells, principally in the lateral geniculate nucleus, causing unstable binocular control resulting in fixation slip and, consequently, all the symptoms of visual dyslexia. Treatments include occlusion and blue or yellow tinted lenses to stimulate or inhibit the magnocellular pathways. Convergence, fixation and tracking effects are reported and many of the symptoms do appear to resolve. Unfortunately much of what is proposed does not match up with the symptoms described and to suggest, that all of the symptoms described by the patient are a binocular anomaly, is clearly nonsense. However, it is likely that magnocellular dysfunction does play a part in the trigeminal and vestibular reflex problems, that are described, as well as convergence, fixation and tracking problems.

Using just blue or yellow lenses shows a lack of understanding of colour space and it was interesting to note that, in the colour space analysis undertaken in the Cambridge trials, not a single person achieved optimum results, by using pure yellow light, and optimum visual results were different, for virtually every patient. I would consider that choosing a colour, without the ability to analyse and assess using the lens, the illuminant and the optimal colour space position, is foolhardy. The only method, that allows the optimum colour space position to be assessed, is the Orthoscopics system.

Occlusion often helps for a number of reasons, not just binocular control difficulties.

Overstimulation

Many people have come to the conclusion that the volume of visual and other sensory stimulation is too great to be accepted by the brain and that a reduction in stimulus input can alleviate symptoms. Whilst a reduction in visual or other sensory stimulus does help in many cases, it does not explain why the majority of the symptoms are present. Treatment is usually to reduce lighting or cover one eye to reduce stimulus levels. This is often directly contrary to the advice, usually given by optical professionals if reading is a problem. They will usually say "a good strong light will help"- don't take their advice, unless they understand perceptual problems!

If simple over-stimulus was the problem, then any reduction in input would be beneficial. This is not the case and it is often a matter of stimulus modification that is required rather than just reduction in input.

Developmental delay

Many professionals (particularly behavioural optometrists) believe that the principal cause of visual dyslexia is that the "normal" development of the visual system has not taken place and that by exercises and visualisation techniques they can either reinstate the developmental pathways or produce new pathways within the system. The assumption, that the plasticity of the system can cope with the long treatments involved and the person will complete the course of treatment, is debatable and reversion is a common problem. Behavioural theories do not explain many of the symptoms experienced by patients, although many behavioural optometrists insist that their technique is the only "correct" method for treatment (this is a claim that is hotly disputed in optical circles). There are also serious cost and time implications, for this method of treatment. A second form of developmental theory is expounded by a small number of optical professionals – syntonics. This suggests, that the sympathetic and parasympathetic systems are affected by colour and that treatment results in dramatic increases in the near visual field, as well as other visual effects. Treatment involves looking at coloured lights for long periods.

Attentional difficulties

There are suggestions that fixation and, consequently, reading difficulties are due to difficulties with filtering out unwanted information from the visual system. This could be due to a magnocellular problem or a processing problem, within the visual system. It does not address many of the symptoms described by patients, although it may have a part to play in the overall picture.

The Dynamic model

One of the reasons for writing this book is to suggest a new model for the symptomology, experienced by ten percent of the population and to show how virtually all symptoms described can be explained and treatments predicted. It will demonstrate that all previous ideas can be brought under a unified model and also why the treatments work. Anatomical structures within the visual system can explain the symptomology and the results expected. Other visual conditions will become easier to understand and new treatments for these may become possible. It is however a speculation, open to amendment or rebuttal, but it is, I believe, significantly better than the theories that are current.

The basic idea

The eye/ brain neurological relationship is a dynamic two way neural network. The anatomical structures reflect this and the processing system requires a "dialogue" between the retina and the cortex that prevents confusion. It has been assumed that all or virtually all of the processing of the retinal information is performed in the visual cortex. However the evidence indicates that some of the processing is at a retinal level and if this processing is inadequate the symptoms of visual dyslexia, amblyopia, strabismus and other conditions are predictable as is their methods of treatment.

The retina

Retinal organisation is I believe at the heart of the difficulties suffered by so many children.
Individual retinal cells may fire one or a number of ganglion cells depending on their position on the retina. Receptive fields at retinal level may be determined by horizontal and / or

amacrine cells and I am going to speculate as to the effects of these cells in visual dyslexia.

We assume that the fields are built up into a unified picture at cortical level but it is clear that these fields have to change their relationships depending on position of gaze as relative magnification between the eyes changes when looking to the side.

These changes in relationship of size could be determined by the horizontal and amacrine cells, depending on position of gaze, if a feed-back loop was in place. Could this be the reason why there is so much feedback from the brain to the retina? **In other words, could it be that the brain is sending back processed information to the retina, thereby modifying the fusional fields, dynamically?**

The wiring of the horizontal and amacrine cells are web-like and would allow changes in fusional fields to take place and disparity could be acceptable within certain limits.

Each individual receptive field has the ability to be either excited or inhibited by light and influence the area around it either by lateral excitation or inhibition. The sequences of the fields would normally be in the same order and if the information provided is desynchronised it would normally be ignored or suppressed.

The small fields would become interrelated because of the cells firing, in more than one field at a time, and larger fields would grow from the smaller fields until a unified vision was built.

There is also a need to integrate the coloured field of each of the cone cell types. If we think of the retina as a three dimensional chessboard with each type of cone cell having a different size square and the system having to integrate not only the overall fields but the coloured fields to allow colour vision to take place. It looks likely that, depending on the field, one of the colours may be dominant in field selection and that, therefore, fusional areas may change depending on colour mix. But, if this field building went wrong or received inadequate stimulation, what would be predicted?

The answer simply is – visual dyslexia.

How symptoms relate to the Dynamic model

There are many symptoms reported in visual dyslexia. They fall broadly into a number of areas.

- Binocular stability difficulties
- Letters or words changing shape
- Letters or words changing orientation
- Letters or words moving to a different position
- Reduction in visual acuity
- Visual field defects
- Vibration of words or letters
- Magnification or minification
- Pattern glare
- Double vision
- Photophobia
- Vestibular difficulties
- Memory difficulties
- Sensory integration problems
- Asthenopia headaches and migraine
- Concentration

I will show, how all the above symptoms can be predicted by fusional area dysfunction, as per the integration model. Virtually every symptom, shown by patients with visual dyslexia, is able to be explained and many other optical conditions may be helped, using perceptual prescribing. However, it is essential that the correct tools are used and the best that I am aware of, are part of the Orthoscopics system.

The symptoms explained

Binocular stability difficulties

Many people with visual dyslexia show convergence difficulties, tracking problems and muscle anomalies, such as exophoria. If fusional areas are not flexible enough, during eye movement, it is inevitable that these anomalies result. In extreme cases, a squint may be found. Using colour, it is possible to modify the retinal relationships, and muscle balance problems show immediate resolution or improvement, in most cases. Even squints can sometimes resolve, immediately, using these non-invasive methods.

Letters or words changing shape
Words or letters change shape, as a consequence of their size, font style and the lighting. Changes may be of individual letters,

words or the complete text. Changes may include letters changing in appearance, to that of other letters or numbers. Changes to the size of the text, the lighting, the font style or the position can immediately resolve the problem. This indicates that the retinal area covered determines the appearance of the text and that the fusional areas are key in producing this symptom.

Letters or words changing orientation

A very common symptom of visual dyslexia is to see either small words e.g. was / saw reversed or see some letters e.g. d/b or p/q reversed. Occasionally the letters may both be seen; for example, a d and a b may appear as a line with two circles attached at either side. Rarely, the words or letters may by turned upside down – often with one eye only. The Optimeyes will allow the person to see them revert during gaze and the visual tracking magnifier will stop the effect, immediately. In other words, changing the size and the colour will stop the symptoms. As it is possible to see the letters changing orientation, it has become apparent that a critical level of both size and illumination will cause the effect to happen and this can be very precise in some cases.

Letters or words moving to a different position

A less common but very disturbing symptom is that of displacement. In this symptom, a letter, word, line or the whole text appears to move to a different position. For a letter, this may be to a different position in the word or to a different word or to a different position on the page. It can easily be mistaken for a sequential memory or phonic problem by the teacher. Sometimes, the letter will disappear, either in the word or at the beginning or end of a word. Again changing colour or size will resolve the problem. These phenomena may be explained by the fusional area combining incorrectly or moving to an area that is suppressed by the system, when an object is of a particular size.

Reduction in visual acuity

An interesting fact is that many people will have a reduction in visual sharpness that can be immediately treated and improved by colour. This indicates that monocular ability may be compromised by inappropriate stimulation. This may provide a better method of improving a lazy eye than using patching techniques although further research needs to be undertaken. It is interesting that many, with reduced acuity, do perceive the image size to be smaller until the correctly prescribed filter is worn and the image size increases.

Visual field defects

There is strong circumstantial evidence that those with reading problems have a reduced functional visual field, that increases with correctly prescribed filters. Experimentally, we have found patients that suffer from a field loss may benefit from correctly prescribed colour appliances, although it is too early to say which patients can benefit most.

Vibration of words or letters

Some patients describe words or letters as "vibrating within words". This could be due to fusional areas being unstable. Again colour or magnification will stabilize.

Magnification or minification

Changes in relative size (may be of dramatic proportions) are occasionally reported. Sometimes, objects will change slowly, but in some cases it can be a rapid change. It may be due to the wrong fusional areas being utilized. Again colour will resolve immediately.

Pattern glare

Text or pattern can create major difficulties for some. Difficulties depend on the contrast, the spaces between the text, the font and the lighting. Again, it is obvious that the fusional areas are crucial in producing symptoms. Pattern glare is a major yet unrecognised problem in schools and offices.

Double vision

Double vision is a common symptom in visual dyslexia. There are two types generally seen in visual dyslexia, one, in convergence, produces two crossed images of the background, the other, that may only be found in one eye, in which only part of the visual field is double. In the first type the sizes of the image vary depending on illumination and degree of convergence. It is apparent, that some colour is more tolerable for individuals. For the second type, the image is dependant on the size, the font, the illumination and the background. It appears that this too is related to retinal organisation. Colour modification will resolve the problem immediately.

Photophobia

A very common symptom of visual dyslexia is sensitivity to light. This is an understandable response if light can cause confusion to the visual system.

Vestibular difficulties

The vestibular system (balance) is highly dependent on the eye and head position. Should fusional areas be at a different relative position there will be a tendency for the head to tilt to allow these areas to coincide and to change, depending on light stimulus. This will have effects on proprioception and the ear position, causing some degree of balance difficulties. Changing illumination will often change posture, balance and responses to head movement.

Memory difficulties

Short term sequential memory can be changed by lighting (positively or negatively). It is likely that this is not a retinal effect and is due to the changes in the alpha waves in the occipital cortex seen in the Cambridge trials.

Sensory integration problems

Auditory effects, such as difficulties with filtering background information and difficulties in distinguishing certain sounds such as "f" and "th", could be due to changes in the hearing system, induced by a trigeminal response or due to the changes in the alpha waves in the occipital cortex. Either of these, will respond to colour treatment.

Asthenopia headaches and migraine

Frontal headaches, dry sore or itchy eyes and migraine respond extremely well to colour or light treatment. They usually suffer a significant level of pattern glare and many of the other symptoms. Migraines can be treated very successfully using the Optimeyes lamp or lenses. The best lenses I have seen for this purpose are the V DEX range.

Concentration

The Cambridge trials demonstrated, conclusively, that lighting can cause major detrimental effects on the neural processing of some subjects. Any child, that does not concentrate at school, should insist on being screened.
If the child cannot either see a word correctly, is stressing the visual system or cannot process the information it is little wonder that they cannot concentrate and look around or out of the window

The Dynamic model explains all the symptoms, commonly reported, but is open to revision and I would be grateful for any constructive comments.

A number of other effects have been reported by some patients to colour treatment. These include mood changes, changes in seeing flicker, changes in Synesthesia, changes in touch sensitivity, changes in sleep patterns…. It is possible to explain all these, by using integration modelling but, apart from the flicker sensitivity, other symptoms are rare.

Colour is a powerful tool, in the right hands, so make sure that you are assessed using the most powerful method, as results can be so much better. For information www.orthoscopics.com

Chapter 11

The future

Light stimulus is the largest sensory input to the brain and about eighty percent of its processing power is devoted to analysing the input signal from the eyes. It is not surprising that, sometimes, this system does not always work perfectly; it is more surprising that it works so well! The system is extremely difficult to investigate and many ideas have to be guessed, as a response to the symptoms exhibited. It is almost impossible to investigate many of the structures, although new imaging methods are going to change methodology, in the future.
One thing is for certain though – things cannot stay the same.

In education, teachers must recognise that visual perception problems are a reality and that the position, that all reading and comprehension difficulties are language based, cannot be maintained. The evidence is now overwhelming and cannot be ignored. Special educational needs teachers must be able to recognise the difference between visual and other problems.

They must be able to screen children using colour; the Optimeyes is the best method. They must be able to use the magnification; the visual tracking magnifier is best. They must be able to refer and understand the report, from the optician that specialises in prescribing for perceptual difficulties. They must be aware of the difference between a simple eye examination and a full perceptual assessment.

Educational psychologists must be aware that the ambient lighting can cause a major change in the results of their assessment and that, unless lighting is controlled, they must consider referral, prior to assessment. They should also be able to understand the report that the optician sends and take it into account.

Any educational needs assessment must include the opticians report, unless it is clear that there is no visual component in the educational problem.

Optical professionals also have to change their working practices. They must be aware of when colour assessment is indicated and either assess the patient themselves, or refer to someone that can assess. Commercial considerations are unacceptable as a reason for non-referral. The public have a right to expect expert assessment.

Medical staff should be aware of the effects of light on a number of conditions and be aware of the relative merits of treatment, using drugs or surgery. Changes are inevitable in treatment of many medical conditions.and research is required for doctors to refer for colour treatment if appropriate..

The public are becoming aware of the potential benefits of light treatment, the tide cannot be turned back and the circle of underachievement must be broken. The way forward is clear – colour can crack the code!

For information on assessments www.orthoscopics.com

Appendix1

Does my child have visual dyslexia? – a check list

This list will give a list of questions and the answers to determine whether your child is likely to suffer from visual dyslexia.

Does your child significantly underachieve at school relative to their ability?

Does your child suffer from headaches in the eye or temple area?

Does your child reverse or invert words or letters?

Does your child see the words "jumble" or look "funny" after a while?

Do the words or letters move or vibrate?

Do the words or letters move to a different part of the page?

Does the child ever get double vision?

Is it hard to copy from the blackboard?

Do words, bits of words or letters disappear or change their appearance?

Is tracking along a line a problem?

Does the child have difficulty in looking at a word or text?

Is the child clumsy?

These twelve questions should be sufficient to screen most cases of visual dyslexia. If your child answers yes to two there is likely to be some sort of visual or visual processing difficulty. The more they answer yes to, the worse they will suffer. Do not delay, they will not grow out of it !

Symptoms check list:

Fixation problems

      Tracking problems

      Reversals of words or letters

      Inversions of words or letters

      Double vision

      Letter or word displacement

      Words or letter vibration

      Word or letter movement

Changes in size or shape of individual letters

Suppression of background

Trigeminal nerve effects

Spatial problems

Visual plane variations

Memory access problems

Memory displacement

Central visual field changes

Peripheral field changes

Proprioceptive problems

Auditory processing difficulties

Synesthesia

Mid line crossing difficulties

Fusional difficulties

Stomach or allergy problems

A history of ear infections

Underachievement

Appendix 2

Amblyopia and strabismus

If retinal de-synchronisation has effects on visual sequencing, does it have other potential effects on the visual system? The answer is a resounding yes!

Although clinical trials have yet to be undertaken to establish parameters, I have demonstrated the effect to a number of optical and other professional meetings. Squint will sometimes resolve spontaneously and muscle balance can be changed for a significant number of people. This effect is disputed by some, it is either present or not – and the Optopraxometer measures it, under controlled conditions! I believe the position of "no effect" is an impossible one to hold, as it directly contradicts the growing body of evidence. Amblyopia improves for some immediately, usually by only a line or two on the optician's chart but sometimes it is much more dramatic.

To understand retinal fusion we have to consider the tristimulus of colour fusion.

Colour perceived is a combination of either excitation or inhibition of the cone cells in a given area. If these cells are synchronised and the fusional areas are synchronised, then normal colour response will be elicited. If, however, this is not the case, then faulty inhibition may be caused resulting in reduced acuity and inability to fuse images that fall on particular areas of the retina. Destabilisation of vision is inevitable, unless suppression takes place. The implications for these suggestions are far reaching and controversial, but the evidence is available and easily reproduced.

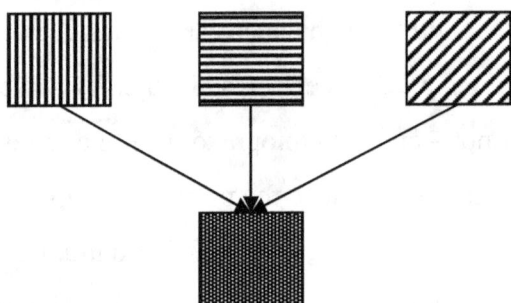

Combination of all types of cone cells response fused into resultant colour to achieve maximum acuity

Appendix 3

Flicker and edges

As we have seen, colour is a combination of wavelengths in the visible spectrum, which together are interpreted as a single colour.

Combination of stimulus

If we imagine a coloured stimulus being received by the visual system, the resultant response is a combination of the responses from the three types of receptor cells (cone cells). This can be considered to be an additive stimulus of different waves. The shape and frequency of the resulting cell responses will determine the tristimulus response, which is closely related to the level of stimulus of each of the different types of cell and how the colour is perceived (metamerism, the phenomenon in which different combinations of wavelengths can actually elicit the same tristimulus response, may also be involved). It also depends on the relative sensitivity of the different cone cells to the amount of light received and it is likely to depend on the firing response to flicker, if it is present. The faster that a particular cone cell can respond to change in stimulus, the more likely that flicker will be observed at the wavelength to which it is sensitive. It has been born out in practice that, often, a different colour prescription is indicated when flicker is present, and also some people can perceive much higher flicker frequencies than others. It explains why colour variations modify the ability to see flicker, why extreme responses to flicker are possible (in particular coloured strobe lights) and the sensitivity to fluorescent lights, experienced by many.

In addition high contrast patterns will increase this effect, as great differences are to be found between the on and off areas (the edges of letters) and lateral inhibition and excitation occur, creating an overload situation. This is the exact situation, which we expect our children to tolerate in the classroom - high degrees of flicker and a large area of high contrast text! The area of the text, the spacing and the contrast all play a part in increasing the problems to the visual system. It has been found that the separation between high contrast areas within patterns on the retina and the pattern design itself create many of the problems. This is likely to be due to retinal structure and, although some may believe it to be cortical, I feel this to be unlikely, as the responses from patients in both monocular and binocular viewing and symptoms related to position of gaze, are size and colour related.

The flicker and edge recognition system is the magnocellular system and it has been postulated that a person with an abnormal magnocellular system will exhibit many of the difficulties described. I suspect that the abnormal cells found are a consequence of abnormal stimulation and development and NOT the cause of the problem.

Colour Prescribing

Treatment, using colour, can have extremely beneficial effects but can also be detrimental. It is essential that all optical professionals should know how to use colour and, at present, this is a rare ability. It is unacceptable for any professional to be unaware of the full effect of many of the lenses that they can prescribe. The optician should prescribe from a position of knowledge; the status quo is unacceptable and untenable. He or she must combine the needs of the subject, the expected lighting conditions and the filter characteristic to give optimum results.

This is the purpose of the Orthoscopics system.

Appendix 4

Results of Cambridge trials

EEGs

Compare these EEGs taken from the Cambridge trials
The alpha waves in the control group can be compared to the
dyslexic group under school lighting conditions. In other
words the dyslexic group were asleep in the class with their
eyes open!

Control group

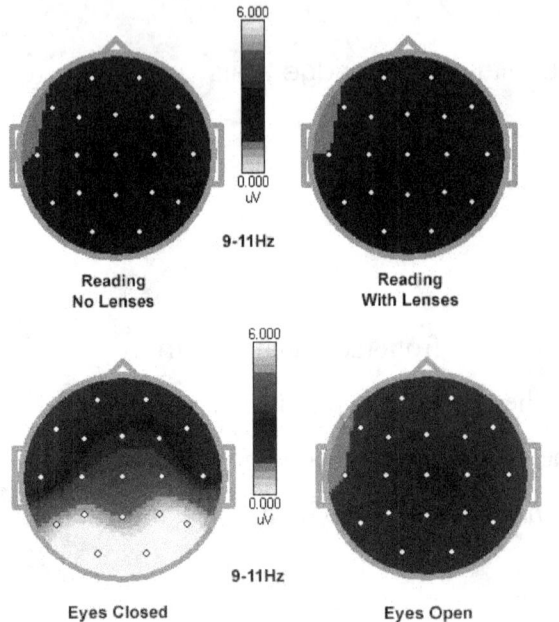

Reading
No Lenses

9-11Hz

Reading
With Lenses

Eyes Closed

9-11Hz

Eyes Open

Dyslexic group

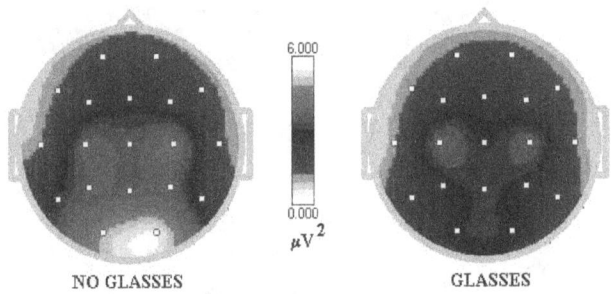

NO GLASSES

μV^2

GLASSES

Compare the results of the optopraxometer tests

Results are consecutive 1 before, 1 after etc.

Test 1-3 Right, mid line, left convergence to spot moving towards viewer.

Test 4 pursuit movement and convergence

Test 5 accommodation

Test 6 central visual stability

Test 7 peripheral stability

Tests are measured in cm from the patient and are the average distance at which difficulties are present

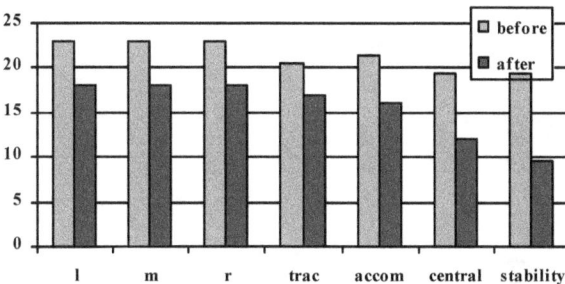

Same tests with cumulative results on mid range dyslexics

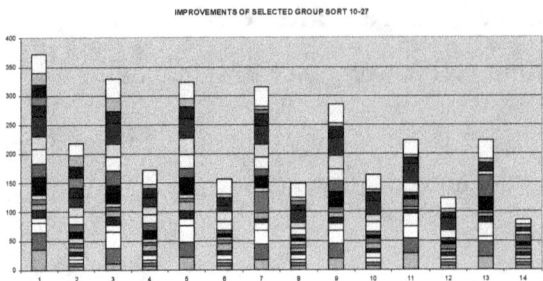

IMPROVEMENTS OF SELECTED GROUP SORT 10-27

The dyslexic group show cumulative distances- note the major improvements in ALL measured parameters

The balance tests averaged out for groups

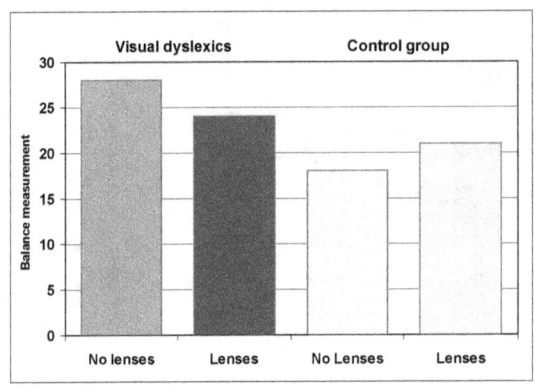

Balance improved with correct prescribing in dyslexic group, deteriorated with randomly prescribed lenses in controls

202

Appendix 5

The cascade of response

If we imagine the response of a single retinal cell to stimulus (the black cell in the centre of the grid) it inhibits the adjacent white cells and produces a secondary level of excitation (the grey cells) which diminishes in intensity as it becomes more distant. This combines with the excitation and inhibition of adjacent cells, to produce an overall level of excitation. The assumption is made that these areas are linear but this is not necessarily the case. If the cell areas build up sequentially, there is a possibility that some blocks of cells are mis-sequenced and this may account for virtually all the symptoms found in visual dyslexia and a number of symptoms in eye disease and dysfunction.

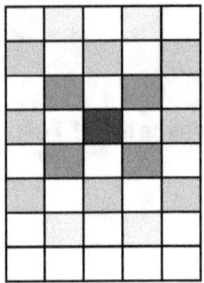

Reversal of letters Reversal of letters

Let us imagine that the sequences of cells have been reversed
in a small area of the retina covering one letter

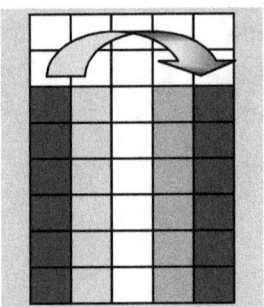

we will consider the letter d

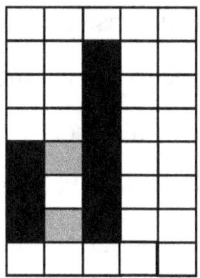

as can be seen the letter is seen reversed by the child

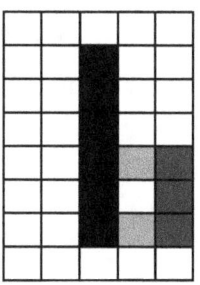

the same effect can be used to show inversion, displacement, disappearance etc. Retinal sequencing may be of serious importance in many conditions and I believe that it cannot be ignored.

It is worth noting that the suggested retinal sequencing mechanism has allowed me to demonstrate some very exciting and predictable effects at conferences such as the British Dyslexia Association International conference but it is a controversial area of practice and many do not as yet acknowledge the phenomenon. (Effects demonstrated include a subject seeing letters reverse in front of his eyes, sequences of letters change, letters disappearing and compressing, eye movement changing, accommodation and convergence modification, the appearance of the page changing, balance and hearing changing, posture and gait changing, sequential memory improving immediately etc..) I measure these effects every day!

Appendix 6

Cranial nerves are they affected?

Symptoms consistent with nerve involvement are reported

These are

Cranial nerve number

I olfactory unlikely or rarely

II optic nerve no obvious effect

III oculomotor accommodation, eye movement, convergence, pupil size (dilation), proprioception, upper lid droop, diplopia and eye movement down and out could be affected.

IV trochlear possible eye movement effects (rotation and direction)

V trigeminal pain and discomfort in facial or temple areas, migraine, epilepsy, hearing, proprioception possible effects

VI abducens unlikely or rarely

VII facial dry eye, dry mouth, blepharospasm

VIII vestibulocochlear balance, vestibular reflexes, nystagmoid movements, tinnitus, vertigo, hearing.

IX, X, XI, XII no reported changes

www.ingramcontent.com/pod-product-compliance
Lightning Source LLC
Chambersburg PA
CBHW081046180526
45170CB00005B/1713